The
World's
Most
Famous
Math
Problem

The World's Most Famous Math Problem

[THE PROOF OF FERMAT'S LAST THEOREM
AND OTHER MATHEMATICAL MYSTERIES]

Marilyn vos Savant

ST. MARTIN'S PRESS NEW YORK

For permission to reprint copyrighted material, grateful acknowledgement is made to the following sources:

The American Association for the Advancement of Science: Excerpts from *Science,* Volume 261, July 2, 1993, © 1993 by the AAAS. Reprinted by permission.

Birkhäuser Boston: Excerpts from *The Mathematical Experience* by Philip J. Davis and Reuben Hersh © 1981 Birkhäuser Boston. Reprinted by permission of Birkhäuser Boston and the authors.

The Chronicle of Higher Education: Excerpts from *The Chronicle of Higher Education,* July 7, 1993, © 1993 *Chronicle of Higher Education.* Reprinted by permission.

The New York Times: Excerpts from *The New York Times,* June 24, 1993, © 1993 *The New York Times.* Reprinted by permission. Excerpts from *The New York Times,* June 29, 1993, © 1993 *The New York Times.* Reprinted by permission.

Cody Pfanstiehl: The poem on the subject of Fermat's last theorem is reprinted courtesy of Cody Pfanstiehl.

Karl Rubin, Ph.D.: The sketch of Dr. Wiles's proof of Fermat's Last Theorem included in the Appendix is reprinted courtesy of Karl Rubin, Ph.D.

Wesley Salmon, Ph.D.: Excerpts from *Zeno's Paradoxes* by Wesley Salmon, editor © 1970. Reprinted by permission of the editor.

Scientific American: Excerpts from "Turing Machines," by John E. Hopcroft, *Scientific American,* May 1984, © 1984 Scientific American, Inc. Reprinted by permission. Excerpts from "Sophie Germain," by Amy Dahan Dalmedico, *Scientific American,* December, 1991 © 1991, Scientific American, Inc. Reprinted by permission. Excerpts from "Fermat's Last Theorem," by Harold M. Edwards, *Scientific American,* October, 1978, © 1978 Scientific American, Inc. Reprinted by permission.

Springer-Verlag: Excerpts from *Thirteen Lectures on Fermat's Last Theorem* by Paulo Ribenboim © 1979 Springer-Verlag. Reprinted by permission of the publisher.

Time, Inc.: Excerpts from *Time* Magazine, July 5, 1993, © 1993 Time, Inc. Reprinted by permission.

Design by Sara Stemen

ISBN 0-312-10657-2

First Edition: November 1993

10 9 8 7 6 5 4 3 2 1

Dedicated to
Isaac Asimov,
with thanks for the memories.

Contents

Acknowledgments

I WANT TO thank the incomparable Martin Gardner for reading my manuscript, for being my dear long-distance friend, and for brightening the world for all thinking people. Author of the Mathematical Games column for *Scientific American* from 1956 to 1981, he wrote the brilliant *The Annotated Alice*—in which he showed how Lewis Carroll's novels actually were structured with chess games, coded messages, and diaphanous caricatures—as well as dozens of other books. I also want to thank Robert Weil, my editor at St. Martin's Press, for having the courage to ask that I write this book in three weeks. If it weren't for him, I wouldn't have known I could do it.

And a personal "thank you" to Barry Mazur, Kenneth Ribet, and Karl Rubin for being such good sports and for putting up with my faxes. Barry Mazur of Harvard University received his Ph.D. from Princeton University and is a member of the U.S. National Academy of Sciences; he has received the Vebien Prize in Geometry and the Cole Prize in Number Theory from the American Mathematical Society. Mazur generously provided me with a text that evolved from his talk at the Symposium on Number Theory, held in Washington, D.C., under the auspices of the Board on Mathematical Sciences of the National Research Council.

Foreword

THIS ISN'T A math book. Instead, it's a book *about* math—
about the possible end of a 350-year-long search and the start of
a new one. Written for the non-mathematician, this book attempts to show the "queen of sciences" in a new way—as a science and as an art. You'll probably learn things about
mathematics that you didn't know before—about its triumphs
and failings, about its human aspects, and about its limits. Regardless of whether the new proof of Fermat's last theorem holds
up under scrutiny, you'll learn how mathematics has thrown off
the yoke of Euclid's legacy and ventured into the deepest waters
of the imagination, whether for better or for worse. And you'll
find it understandable to read, regardless of the extent of your
mathematical education.

Karl Rubin of Ohio State University, who received his Ph.D.
from Harvard University under the supervision of Andrew Wiles,
and who was in attendance at the meeting where Wiles presented

his proof of Fermat's last theorem, has generously allowed us to reprint his brief sketch of the highlights. (Rubin is best known for his work on elliptic curves, a special class of equations that play a fundamental role in the proof; he received the Cole Prize in Number Theory in 1992.) The sketch was sent through electronic mail to his math newsgroup following the last lecture, and we've left it intact in the appendix with all of its "you are there" charm, including the e-mail salutation and the translation idiosyncrasies, such as ^ instead of superscript. You'll find a not-very-plain-English version of the proof in the text.

The
World's
Most
Famous
Math
Problem

One of the Hottest Stories in the History of Math

[*Archimedes Thinking in the Bathtub*]

"EUREKA [I HAVE FOUND IT!] Then again, maybe I haven't."
It's no wonder that Archimedes didn't say that second sentence
while sinking down into a warm bath. In the legendary story from
the third century B.C., the Greek tyrant Hiero II asked the re-
nowed mathematician and physicist to find a method for determining whether a crown was made of pure gold or alloyed with
silver instead. Considering Hiero's notoriously unpleasant tem-
perament, Archimedes was lucky to realize, as he stepped into
the tub, that a given weight of gold would displace less water than
an equal weight of silver. (Gold is denser than silver, so a gold
coin would be smaller than a silver coin that weighed the same.)
In the throes of mathematical ecstasy over this momentous discovery, he is supposed to have run home stark naked, shouting
"Eureka! Eureka!" ["I have found it! I have found it."].

Archimedes was also lucky to be given a task that would yield
to instant insight. (It's a good thing that showers—that "colonial

abomination"—weren't popular then; mathematical progress might have been considerably slowed.) Proving Fermat's last theorem undoubtedly would have taken him considerably longer. But as of June of 1933, that puzzle has at last been solved. Well, maybe.

[*Pierre de Fermat Writing in the Margin*]

More than 350 years ago, a French mathematician and physicist named Pierre de Fermat wrote down an apparently simple little theorem in the margins of a mathematical book he was reading. The theorem was about what solutions are not possible for certain elementary equations. Fermat added that he had discovered a remarkable proof for his statement, but that there was no room in the margin to include it. He died without ever presenting the proof to substantiate this tantalizing claim. The best of mathematicians have been trying to do so ever since.

As Nigel Hawkes, Science Editor for *The Times of London*, put it on June 24, 1993, "Since Fermat wrote down the theorem, it has become the mathematical equivalent of Schubert's *Unfinished Symphony*." Two days later, the same newspaper labeled a proof of the theorem "the mathematical result of the century," calling it "as spectacular in its field as the discovery of Shakespeare's alleged *Love's Labour's Found* or the authentication of a lost Botticelli."

It's not as if no progress has been made at all. Since the arrival of computers, the theorem clearly has been shown to hold true for extremely high numbers. In 1992, an enormous computer effort verified Fermat's last theorem for exponents up to four million.

That might seem proof enough for the general public, but for mathematicians, it's no proof at all. Many of them were coming to the reluctant conclusion that Fermat's reach may have exceeded his grasp at last. "It has always been my belief that Fermat made a mistake," said Dr. Harold Edwards, a professor of mathematics at the Courant Institute of Mathematical Sciences at New York University and author of *Fermat's Last Theorem*, the definitive book on the subject, as quoted in *The Chronicle of Higher Education*. Still, mathematicians kept searching for the elusive proof.

Then, on July 2, 1993, came the news that sent shock waves through the mathematical community around the world. After dozens of claims of success made too early, hundreds made too unwisely, and thousands made too amateurishly, many mathematicians are cautiously heralding the work of a publically obscure, though highly respected colleague. They believe he may have at long last conquered what is perhaps the most intimidating test of strength and endurance and probably the most famous of unsolved problems in their corner of the intellectual landscape. (Mathematical details of Fermat's last theorem and an overview of Andrew Wiles's proposed proof can be found in Chapter Two. An abbreviated sketch of the proof itself can be found in the Appendix to this book.)

[*Andrew Wiles Lecturing at Cambridge University*]

Dr. Andrew Wiles is a reserved, bespectacled 40-year-old English mathematician at Princeton University. (It goes without saying that he's also very intelligent; there are no mathematicians who *aren't*.) Wiles became the subject of intense interest when he made a modest announcement to about seventy-five mathematicians at the end of a three-day lecture series at the newly opened Isaac Newton Institute for Mathematical Sciences at Cambridge University, in England, where he had done his doctoral studies.

This is how *Science* magazine decribed the atmosphere as the lectures were about to begin: "He was equally quiet when he arrived at the Newton Institute . . . but rumors of a breakthrough were starting to fly among the other participants—in part because Wiles, who normally doesn't ask to give lectures, had asked to give not just one, but three hour-long talks. John Coates of Cambridge University, who was Wiles's thesis adviser at Cambridge in the mid-1970s, scheduled him on Monday, Tuesday, and Wednesday, June 21–23.''

Time magazine reported on the growing excitement in the hall: "By the end of the first hour . . . , they knew something was up. Recalls Nigel Boston, a visiting mathematician [there]: 'We realized where he could be heading. People were giving each other wide-eyed looks.' By the end of the third hour, the room was packed with excited number theorists. Wiles finished up his talk and wrote a simple equation on the blackboard, a mathematical afterthought that logically followed from all that he had been saying.''

According to *The Associated Press* report on the following day, "Dr. Peter Goddard, the institute's deputy director, said . . . that nobody knew [Wiles] had come up with a solution until the very end of a lecture in which he had written down many mathematical results. 'He wrote down the last line, and the last line was a corollary to his last result, and it was Fermat's last theorem,' Goddard said. 'Then, he turned to the audience and smiled and said something like, 'I better stop there.' ''

Reuter's reported, "There was a moment of stunned silence followed by rapturous applause as the enormity [sic] of the event sank in." Telephones began to ring, faxes churned out copy, electronic mail zapped into computers all over the world, and the communications satellites went into overdrive.

Could it really be true? Mathematicians were first startled, then excited, and ultimately—well, the emotions are mixed. One day after the event, *The New York Times* quoted Dr. Leonard Adelman of the University of Southern California as saying, "It's the most exciting thing that's happened in—geez—

maybe ever, in mathematics." But on June 29, 1993, the paper ran a more reflective response from Wiles himself.

> There is a certain sadness in solving the last theorem. All number theorists, deep down, feel that. For many of us, his problem drew us in, and we always considered it something you dream about, but never actually do. There is a sense of loss, actually.

[*The Five-Point Test*]

Surely, a great many mathematicians must share Wiles's sense of loss. But, just as surely, many may lack his concurrent elation. For them—and also for Wiles, of course—there are several further observations that merit consideration. The following is my own general guide to the analysis and ultimate evaluation of a proof. I call it the "Five-Point Test."

[1] *Might there be a subtle error in the proof?*
  Yes. The two-hundred-page proof is unpublished as yet. Until it appears in mathematical journals, which could take months, and until it is thoroughly checked and rechecked and double-checked, there remains the possibility that it contains a significant error. (There are also likely to be a few minor errors that can be corrected without damaging the overall proof. This is called "polishing" the proof.)

[2] *Does the proof rest on numerous other proofs (especially recent, arcane, obscure, or delicate ones)?*
  Yes. In 1954, the late Japanese mathematician Dr. Yutaka Taniyama made a conjecture about elliptic curves (a particular class of cubic equations). Taniyama's conjecture (known also as Taniyama-Weil for broader attribution) was further clarified by Dr. Goro Shimura of Princeton University. (The conjecture now can be known as Shimura-Taniyama-Weil, but for simplicity's sake, further references

will be to Taniyama alone, as is most common elsewhere.) No one, however, thought that this conjecture had any relationship with Fermat's last theorem. Now the plot thickens. About ten years ago, Dr. Gerhard Frey of the University of the Saarland in Germany postulated a connection between Taniyama's conjecture and the theorem. In essence, it suggested that if the conjecture were proved, so would be Fermat's last theorem.

Dr. Kenneth Ribet of the University of California at Berkeley presented a proof of the connection in 1987. (More specifically, he demonstrated that an elliptic curve can be correlated with a solution to those equations that Fermat maintained were impossible. If it could be proved that such an elliptic curve is impossible, it would follow that solutions to Fermat's equations were impossible, too.) And now, Wiles apparently has proved one form of Taniyama's conjecture, implying that Fermat's last theorem is true (that solutions to certain equations are not possible), at least indirectly.

[3] *Does the proof rest on a whole new philosophy?*

Yes. On June 29, 1993, *The New York Times* quoted Ribet.

Ribet said that the Taniyama conjecture was part of a sort of grand unified theory of mathematics. "The Taniyama conjecture is part of the vast Langlands philosophy," Ribet said. He explained that Dr. Robert Langlands, a professor at the Institute for Advanced Study in Princeton, suggested years ago that two apparently disparate fields of mathematics were actually one and the same. . . . The Langlands philosophy, Ribet said, "is a deep, far-reaching vision in mathematics." The Taniyama conjecture, he added, "is a special case of what Langlands suspected."

[4] *Is the proof unintelligible to most experts in the field?*

Yes, at least so far. However, Wiles is a mathematician of significant stature, and he has a reputation for being extremely careful. Mathematicians are further convinced by the approval of Dr. Barry Mazur of Harvard University and the overwhelming endorsement of Ribet, who was the mathematician instrumental in providing the step that linked the conjecture to the theorem and ultimately to Wiles'[s] proof.

"Wiles'[s] arguments are based on the most advanced, most elaborate mathematics that exist in this field," Ribet was quoted in *Time* magazine, referring to his own work [among others]—upon which Wiles'[s] proof rests. "The number of mathematicians who can really fully understand the arguments would fit into a conference room." And Ribet estimated to *The New York Times* "that a tenth of one percent of mathematicians could understand Wiles'[s] work because the mathematics is so technical."

[5] *Are there any "signs of forced entry," as a detective might call it?*

Yes. According to *The New York Times*, "Ribet's proof linking the conjecture to Fermat's last theorem fired Wiles'[s] imagination. . . . The very day that Wiles heard about Ribet's result, he dedicated his life to using it to prove Fermat's last theorem."

These are the points to ponder, not just with this proof, but with them all. (The "point to ponder" regarding this proof in particular follows shortly.) It shouldn't be surprising, by the way, and shouldn't be regarded as a negative assessment, that the "Five-Point Test" rang cautionary bells on every question regarding Wiles's work. Considering that his apparent achievement is a proof that has eluded mathematicians for centuries, this caution, about too quickly accepting it, is understandable. A proof that rings no cautionary bells is likely to be much more straightforward (and shorter) than one that comes at the end of a three-day lecture series in Cambridge, entitled "Modular Forms, Elliptic Curves, and Galois Representations," part of a conference on "*P*-adic Galois Representations, Iwasawa Theory, and the Tamagawa Numbers of Motives."

However, when a proof is supported by a small group of people, and when others in the field believe that virtually no one outside this group is capable of understanding it, the cautionary bell on Point Four rings a little louder. Who is in an appropriate position to verify the proof? Mathematicians are an extremely bright class of professionals, and most of them should be more than capable of understanding the proof intellectually. The problem is that Wiles's field is so narrowly specialized that few mathematicians have the time it would take to educate themselves in that particular area.

[Dr. Wiles's Experience]

Wiles' credentials are impeccable (he was educated at Oxford and Cambridge Universities and has been a professor at Princeton since 1982; earlier appointments were at Harvard University and the Institute for Advanced Study in Princeton), and he has acquitted himself admirably throughout his professional life.

For one thing, his fellow mathematicians report that he has expressed great reservations about speaking to the press. (This may in itself be evidence of good judgment, considering the mayhem that can result from reckless claims. The cold fusion story, which proved unfounded, is one example.

But then again, perhaps Wiles simply has a keen sense of humor. After all, without a direct explanation from him, journalists all over the world are left to struggle to explain to baffled readers what has mystified even the mathematical community for centuries.) He did tell *The Chronicle of Higher Education:* "I think it's very important that people are encouraged to work on very hard problems. The tendency today is to work on short and immediate problems." In addition to his other qualities, Wiles clearly is possessed of an admirable determination.

His is a classic story, in a way, and if his proof stands the test of time, Wiles will take his place in the history books. (Of course, the test of time often isn't the best test, as anyone familiar with the Ptolemaic universe can attest. The Ptolemaic system was the most important of the geocentric cosmological theories, that is, those theories maintaining that the Earth stood motionless at the center of the universe with all heavenly bodies revolving around it. This theory dominated astronomy for millennia, until the arrival of the heliocentric, or sun-centered, Copernican system in the sixteenth century. In more modern times, Albert Einstein's general theory of relativity maintains enormous influence, but it is unknown whether he eventually will take a place in the history books closer to Ptolemy or closer to Copernicus.)

Born in Cambridge, Andrew Wiles first learned about Fermat's last theorem when he was only ten years old, a fine age at which to develop a lifelong obsession, and he surely did. It was, he reports, the reason he decided to become a mathematician—he wanted to find the solution to the problem. ("As a child, I used to think that Fermat had solved it," he told *The Chronicle of Higher Education*. "My guess now is that if he really thought he solved it, he would have written the proof somewhere.") Not surprisingly, Wiles worked on it with naïve enthusiasm as a teenager, but it wasn't until 1987 (when Ribet presented a proof of the connection between Taniyama's conjecture and Fermat's last theorem) that he took it upon himself as a personal (and private) challenge. " 'I have a preference for working on things that nobody else wants to or that nobody thinks they can solve,' he explains. 'I prefer to compete with nature rather than be part of something fashionable (quoted in the September 1993 issue of *Scientific American*).' "

Very few people knew what Wiles was doing in his little

third-floor attic office at home for seven years, and he wanted it that way for good reason. After all, what would people think? Worse, what would they think if he worked on it for a lifetime and *failed?* The assessments would probably not be charitable, especially for a man with a wife and children and a house with an average assortment of squeaky screen doors, leaf-filled gutters, and dandelions in the backyard. According to the September 1993 issue of *Scientific American*, ". . . Wiles virtually stopped writing papers, attending conferences or even reading anything unrelated to his goal." It would surprise no one if, after seven years of this, every window in the house eventually became stuck shut. (Even so, followers of Fermat certainly knew what he was up to. In Dorian Goldfeld's preface to the proceedings of the conference "Number Theory Related to Fermat's Last Theorem," sponsored by the Vaughn Foundation in May of 1982, it was noted that Harold Edwards, Nicholas Katz, Neal Koblitz, Barry Mazur, and Andrew Wiles himself were the other organizers of the conference.)

Unlike certain scientists who put out self-congratulatory press releases every time they fill up a notebook, Wiles didn't mention beforehand that he intended to do the mathematician's equivalent of lobbing a hand grenade into a fireworks factory. Instead, as the *The New York Times* reported, people were concerned about "whether the proof would be accepted by the handful of mathematicians who understood the field." They need not have worried, at least for the time being. This is how *The New York Times* put it in its June 29, 1993 issue:

> Around 5:30 Wednesday morning, Dr. [John] Conway [of Princeton University] unlocked the door to Fine Hall, the brown tower that houses the mathematics department, went into his office, and turned on his computer. The first message arrived at 5:53 A.M. from Dr. John McKay of Concordia University in Montreal, who was attending the meeting. "F.L.T. proved by Wiles," it said.
>
> Q.E.D.

(Ribet was the next lecturer on the program. It must have been a tough act to follow.) Apparently, Wiles didn't fully expect the reaction he got. "Friends said when he returned to Princeton last week that he had been a bit overwhelmed by the stir he had created," noted *The Chronicle of Higher Education*. It also quoted Simon Kochen, the chairman of the mathematics department at Princeton University. "He finally got an answering machine at home to answer all of the calls." The September 1993 issue of *Scientific American* noted that "Wiles does think his . . . proof can be simplified. Will Wiles take on this task? 'I'm afraid I've made this so fashionable that I may have to move on to something else,' he replies."

[*A n o t h e r   P o i n t   t o   P o n d e r*]

This is the "point to ponder" I mentioned earlier. The mathematics of today is a far cry from the mathematics of Fermat's time, which no one will dispute. The same is true for the field of medicine, and we're all very thankful for that. But I also feel safe in opining that mathematical purists, transported by time machine to the modern era, would be incensed by the practice of mathematics today, which may be threatening to turn what Carl Friedrich Gauss (the German mathematician, physicist, and astronomer) called the "queen of sciences" into an art form, instead. Here's what Philip Davis and Reuben Hersh had to say about the subject in their book *The Mathematical Experience*, winner of the 1983 American Book Award.

The Degradation of the Geometric Consciousness

It has often been remarked over the past century and a half that there has been a steady and progressive degradation of the geometric and kinesthetic elements of mathematical instruction and research. During this period, the formal, the symbolic, the verbal, and the analytic elements have prospered greatly.

What are some of the reasons for this decline? A number of explanations come to mind:

1. The tremendous impact of Descartes' *La Geometrie,* wherein geometry was reduced to algebra.

2. The impact in the late nineteenth century of Felix Klein's program of unifying geometries by group theory.

3. The collapse, in the early nineteenth century, of the view derived largely from limited sense experience that the geometry of Euclid has *a priori* truth for the universe, that it is *the* model for physical space.

4. The incompleteness of the logical structure of classical Euclidean geometry as discovered in the nineteenth century and as corrected by Hilbert and others.

5. The limitations of two or three physical dimensions which form the natural backdrop for visual geometry.

6. The discovery of non-Euclidean geometries. This is related to the limitations of the visual ground field over which visual geometry is built, as opposed to the great generality that is possible when geometry is algebraized and abstracted (non-Euclidean geometries, complex geometries, finite geometries, linear algebra, metric spaces, etc.).

7. The limitations of the eye in its perception of mathematical "truths" (continuous, nondifferentiable functions; optical illusions; suggestive, but misleading special cases).

In brief, here's what's happened in the field of mathematics and how this relates specifically to the current work on Fermat's last theorem. The geometry of Fermat's day was a set of principles that were derived by rigorous logical steps from the axioms detailed by Euclid (hence known as Euclidean geometry), the Greek mathematician of the third century B.C., in his *Elements* (a series of thirteen books covering that subject and much more in mathematics, including a deductive system of proof). These are the first five axioms.

[1] Given two points, there is a line that joins them.
[2] A line can be prolonged indefinitely.
[3] A circle can be constructed when its center, and a point on it, are given.
[4] All right angles are equal.
[5] If a straight line falling on two straight lines makes the interior angles on the same side less than two right angles, the two straight lines, if produced indefinitely, meet on that side on which the angles are less than the two right angles.

That fifth axiom is known "Euclid's parallel postulate" and can be rephrased this way: If a point lies outside a straight line, one (and only one) straight line can be drawn through that point that will be parallel to the first line.

[*Euclidean Versus the New Non-Euclidean Geometries*]

Some mathematicians in the nineteenth century began to disagree with the "parallel postulate." This gave rise to non-Euclidean geometries, of which there are two important forms, both of which replace the fifth postulate with alternatives. (Agreeing with Euclid's fifth postulate came to be known as "taking the Fifth.") One of the two main alternatives allows an *infinite* number of parallels through any outside point. "Hyperbolic" (Lobachevskian) geometry developed from this approach. The other main alternative allows *no* parallels through any outside point. "Elliptic" (Riemannian) geometry developed from this form. Superficially, these definitions of parallels seem ridiculous to the non-mathematician, but the new systems of geometry have their own definitions and systems of logic. (No system of geometry is found in nature, including the Euclidean system. The earth is not a perfect sphere, and cubes don't grow on trees.)

The outcome in both Euclidean geometry and the alternatives is similar—except for the postulates involving parallel lines. (Other postulates are modified.) However, the ramifications of the differences are great regarding subjects like space and infinity. (For example, in Euclidean geometry, the length of a line is infinite; in elliptic geometry, it is strictly finite. In this non-Euclidean universe, space is finite, but unbounded, and it curves back around on itself.)

Jeremy Gray discusses the subject at length in the second edition of his book called *Ideas of Space:* "Euclidean, Non-Euclidean, and Relativistic," noting:

> Aristotle discussed whether thinking that parallels meet is a geometric or an ungeometric error, that is, whether the contradiction which arises from denying the existence of parallels is strictly mathematical or more broadly logical in its nature.
>
> We might discuss a misapplication of statistics in the same way: is a man wrong to apply statistics in this way or to analyze the problem in this way?

For practical applications in the "real world" (such as in engineering, for example), one works only with Euclidean geometry. But in the world of the imagination, one can work with anything at all. This is how Morris Kline stated it in his book *Mathematics for the Nonmathematician.*

> Remarkable and revolutionary developments of another kind also took place in the nineteenth century, and these resulted from a re-examination of elementary mathematics. The most profound in its intellectual significance was the creation of non-Euclidean geometry by Gauss. His dis-

covery had both tantalizing and disturbing implications: tantalizing in that this new field contained entirely new geometries based on axioms which differ from Euclid's, and disturbing in that it shattered man's firmest conviction, namely that mathematics is a body of truths. With the truth of mathematics undermined, realms of philosophy, science, and even some religious beliefs went up in smoke. So shocking were the implications that even mathematicians refused to take non-Euclidean geometry seriously until the theory of relativity forced them to face the full significance of the creation.

It's not that Euclid had been thought infallible. Rather, when non-Euclidean geometry was first conceived, "It seemed to be at the edge of madness," as Philip Davis and Reuben Hersh put it in their book *The Mathematical Experience.* Indeed, Sir Arthur Eddington sounded like the high priest of a mathematical illuminati in his non-Euclidean "bible" *The Expanding Universe,* which often reads like a religious text.

I see our spherical universe like a bubble in four dimensions; length, breadth, and thickness, all lie in the skin of the bubble. Can I picture this bubble rotating? Why, of course I can. I fix on one direction in the four dimensions as axis, and I see the other three dimensions whirling around it. Perhaps I never actually see more than two at a time; but thought flits rapidly from one pair to another, so that all three seem to be hard at it. Can *you* picture it like that? If you fail, it is just as well. For we know by analysis that a bubble in four dimensions does not rotate that way at all. Three dimensions cannot spin round a fourth. They must rotate two round two; that is to say, the bubble does not rotate about a line axis but about a plane. I know that that is true; but I cannot visualise it.

. . . The material system, like the space, exhibits *closure;* so that no galaxy is more central than another, and none

can be said to be at the outside. Such a distribution is at
first sight inconceivable, but that is because we try to con-
ceive it in flat space.

. . . Apart from our reluctance to tackle a difficult and un-
familiar conception, the only thing that can be urged
against spherical space is that more than twenty centuries
ago a certain Greek published a set of axioms which (infer-
entially) stated that spherical space is impossible.

The best-known example of a non-Euclidean idea is Einstein's general theory of relativity, which has little validity
outside elliptic geometry. (Space remains Euclidean in special
relativity). Curved space is fundamental in relativity theory.
Albert Michelson, the first American to win the Nobel Prize
in physics, was quoted in R. S. Shankland's book *Conversa-
tions With Albert Einstein*. (Michelson and Edward Morley con-
ducted the Michelson-Morley experiment, which led to the
refutation of the ether hypothesis—the medium that was supposed to be required for the transmission of electromagnetic
waves in free space—and was later explained by Einstein's
theory of relativity.) Michelson told Einstein "he [Michelson]
was sorry his own work may have helped to start this 'monster'
[the theory of relativity]."

If any part of Wiles's proof, or any of the steps leading up
to it (including, among others, Ribet's proof and the Frey paper on the links between stable elliptic curves and certain diophantine equations), has any non-Euclidean component that
is invalid in Euclidean geometry, that proof inhabits a very
different world from the world inhabited by Fermat. Indeed,
the chain of proof is solidly based in hyperbolic (Lobachevskian) geometry, which Nikolai Lobachevsky himself named
"imaginary geometry" in his esteemed paper called "On the
Foundations of Geometry," which details the complete development of hyperbolic geometry.

"There are many others whose work Wiles had to use,"

Simon Kochen, Princeton's mathematics department chairman, pointed out to the Associated Press. "He was throwing the kitchen sink at it, using all kinds of techniques that had been developed in recent years."

[*Doubling Cubes, Trisecting Angles, and Squaring Circles*]

Three of the oldest problems in mathematics—all more than two thousand years old—are known as "Doubling the Cube," "Trisecting the Angle," and "Squaring the Circle." All constructions were to be accomplished using only a ruler—as a straight edge, not as a measuring device—and a compass. The problem of doubling the cube is to construct a cube with twice the volume of a given cube; the problem of trisecting an arbitrary angle is to construct a method to divide any given angle into three equal parts (it must work for *every* angle); the problem of squaring the circle is to construct a square with an area equal to that of a given circle. These problems have fascinated professional and amateur mathematicians alike, so much so that Arthur Jones, Sidney Morris, and Kenneth Pearson recount in their book about the subject, *Abstract Algebra and Famous Impossibilities:*

> In the time of the Greeks, a special word was used to describe people who tried to [square the circle]—tetragonidzein—which means "to occupy oneself with the quadrature." . . . In 1775, the Paris Academy found it necessary to protect its officials from wasting their time and energy examining purported solutions of these problems by amateur mathematicians. It passed a resolution . . . that no more solutions were to be examined of the problems of doubling the cube, trisecting an arbitrary angle, and squaring the circle, and that the same resolution should apply to machines for exhibiting perpetual motion.

It wasn't until the nineteenth century that all three problems were proven impossible to solve. This was the subject of Jones, Morris, and Pearson's book, which is an abstract algebra textbook. (In its issue of October 1978, *Scientific American* noted that, "Fermat's last theorem differs from circle squaring and angle trisecting in that those tasks are known to be impossible, and so any purported solutions can be rejected out of hand.") Bearing all this in mind, note what Jeremy Gray said in his book called *Ideas of Space* about Jánosor (János) Bolyai, one of the three founders of hyperbolic geometry.

> Finally, he concluded by solving in non-Euclidean geometry one of the classical problems of geometry: SQUARING THE CIRCLE [capitalization added for emphasis] i.e., constructing a square equal in area to a given circle. At the time, it was not known whether this could be done in Euclidean geometry, the first conclusive proof of impossibility being given in 1882. . . . Bolyai indeed pointed out that his proof will not work in the case of Euclidean geometry.

So one of the founders of hyperbolic geometry (the geometry of the current proof of Fermat's last theorem) managed to square the circle?! Then why is it known as such a famous impossibility? Has the circle been squared, or has it not? That issue of *Scientific American* noted that squaring the circle is "known to be impossible, and so any purported solutions can be rejected out of hand." So has Fermat's last theorem been proved, or has it not? *That is, if we reject a hyperbolic method of squaring the circle, we should also reject a hyperbolic proof of Fermat's last theorem!*

This is not a matter of merely changing the rules (for example, using a ruler instead of a straight edge.) It's much more significant than that. Instead, it's a matter of changing whole definitions, including the definition of what constitutes a contradiction. And regardless, it is logically inconsistent to *reject* a hyperbolic method of squaring the circle and *accept* a hyperbolic method of proving F.L.T.!

The next thing you know, someone will use non-Euclidean geometry to prove Euclid's parallel postulate! (And then what a fix Einstein will be in.)

2

Pierre de Fermat and the Last Theorem

[*Pierre de Fermat*]

PIERRE DE FERMAT (pronounced fehr-MAH), the French mathematician who was born in Beaumont-de-Lomagne, Languedoc, on August 20, 1601, and died at Castres, near Toulouse, on January 12, 1665, is now far more famous than he ever was—mainly because he had a habit of scribbling little notes in the margins of the books he was reading. (In Eric Temple Bell's book *The Last Problem*, Bell wrote that Fermat was accused of "proposing problems to the English mathematicians which he had not solved and could not solve," and *The Times of London* called Fermat "a vain and tiresome joker as well as a brilliant mathematician," but it may be best to put that "joker" possibility out of our minds.)

Fermat was married. He and his wife had five children, and they led an uneventful, quiet life close to home, spending vacations at their country house. Along with his contemporary Blaise Pascal, the French mathematician and physicist, Fermat founded the theory of probability. He also studied the properties of natural numbers (the set of numbers beginning with 1, 2, 3, . . . that are used for counting) and was the first to progress beyond the work of Diophantus, the Greek mathematician from the third century. In fact, Fermat is known as the founder of modern number theory.

And it was in number theory that Fermat made his greatest mark—literally. While reading his Latin translation of Diophantus's Greek masterpiece *Arithmetica*, he wrote a deceptively simple comment in Latin next to a problem about finding squares that are sums of two other squares (for example, $3^2 + 4^2 = 5^2$). The comment he made is now known as Fermat's Last Theorem ("F.L.T." to its closest friends). *Scientific American* has aptly translated it into English:

> On the other hand, it is impossible for a cube to be the sum of two cubes, a fourth power to be the sum of two fourth powers, or in general for any number that is a power greater than the second to be the sum of two like powers. I have discovered a truly marvelous demonstration of this proposition that this margin is too narrow to contain.

[*Fermat's Last Theorem*]

Restated, the equation $x^n + y^n = z^n$, in which n is an integer greater than 2, has no solution in positive integers. This means that there are no positive whole numbers that solve the equation when the exponent n is greater than 2. When n is 2, there is an infinity of solutions. One such solution is the Pythagorean theorem, which states that the sum of the squares of the lengths of two sides of a right-angled triangle is equal to the square of the length of the hypotenuse (the side that is opposite the right angle). As an example, $3^2 + 4^2 = 5^2$, or 9 + 16 = 25. (When the exponent is 2, the solution is called a "Pythagorean triple.") Fermat's last theorem means that it is impossible to find any positive whole numbers for x, y, and z when n is 3 or more.

It is uncomplicated to prove that $x^4 + y^4 = z^4$ is impossible; therefore, the original equation is impossible whenever n is divisible by 4. (If $n = 4k$, then $x^n + y^n = z^n$ implies the impossible $X^4 + Y^4 = Z^4$ in which $X = x^k$, $Y = y^k$, and $Z = z^k$.) Also, if $x^m + y^m = z^m$ can be proved to be impossible for any particular m, it follows that the original equation is impossible for any n that is divisible by m. And because every n greater than 2 is divisible either by 4 or by an odd prime number, Fermat's last theorem can be proved in its entirety if it is proved in the cases where n is a prime number.

[*Learned Attempts at Proof*]

Unfortunately, Fermat died without ever offering a proof to the world of mathematics, and people have been searching for one ever since. (Interestingly, this was not the only cryptic note Fermat had ever made in a margin. He made another note in the same copy of *Arithmetica* about binomial coefficients—the factor multiplying the variables in a binomial expansion, for example, in the equation $(x + y)^2 = x^2 + 2xy + y^2$, the binomial coefficients are 1, 2, and 1—which read, according to André Weil in his book *Number Theory*, "I have no time, nor space enough, for writing down the proof in this margin." It's very much like the notorious "last" theorem, but it has already been proved true. In fact, Fermat made extensive marginal notes in that particular volume, and after his death, one of his sons published a new edition of *Arithmetica*, complete with Fermat's marginal notes. The original copy has been lost.)

Over the centuries, the theorem has been proved for individual exponents. Leonhard Euler, the Swiss mathematician, proved F.L.T. was true when the exponent was 3. Fermat himself had proved it was true when the exponent was 4, and so did Bernard Frénicle de Bessy. Following that, Adrien Marie Legendre, the French mathematician, proved it was true when the exponent was 5. It also was proved true for 7 by Gabriel Lamé, another French mathematician, who thought he had discovered a method for proving other numbers. *Scientific American* recounted that story: "The unfortunate Lamé was so carried away by his own optimism that he announced to a meeting of the French Academy of Sciences that he had proved Fermat's last theorem by this method. As soon as he had presented a sketch of his proof, however, Joseph Liouville rose to object . . . There were other weak points, however. Lamé's enthusiasm was so extreme that he overlooked several other serious difficulties. . . . Naturally Lamé was embarrassed by the foolishness of his errors and by having them published in the proceedings of the French Academy for the entire mathematical world to see. 'If only you had been in Paris or I had been in Berlin,' he wrote to his friend . . . in Berlin, 'all of this would not have happened.'" In summary, since Fermat's death, isolated cases have been proved true, but none of these constituted a *general* proof.

Sophie Germain, a French mathematician, made significant progress, proving that the theorem was true under certain conditions for any prime number under 100. (Incidentally, she performed this work in nearly impossible circumstances. Without a formal education—because she was a woman—she taught herself Latin, Greek, and mathematics at home; she was forced to work using a male pseudonym as a disguise.) *Scientific American* elaborated:

> In 1808, Germain wrote to Gauss, describing what would be her most brilliant work in number theory. Germain proved that if x, y, and z are integers, and if $x^5 - y^5 = z^5$, then either x, y, or z must be divisible by 5. Germain's theorem is a major step toward proving Fermat's last theorem for the case when n equals 5.

Gauss never commented on Germain's theorem. He had recently become professor of astronomy at the University of Göttingen, and he set aside his work in number theory. He became consumed with professional and personal problems.

For the most part, Germain's theorem remained unknown. In 1823 Legendre mentions it in a paper in which he describes his proof of Fermat's last theorem for the case where n is 5. . . . Germain's theorem was the most important result related to Fermat's last theorem from 1738, until the contributions of Ernst Kummer in 1840.

Ernst Eduard Kummer, the German mathematician, had developed a new theory of ideal factorization, and he used it to prove the theorem true for all prime exponents below 100 except for 37, 59, and 67.

In 1815 and 1860, the French Academy of Sciences offered a gold medal and three hundred francs to anyone who could prove F.L.T., and the Geman mathematician Carl Louis Ferdinand von Lindemann succumbed to the temptation. After more than five years of work, von Lindemann claimed credit and published a lengthy paper of substantiation in 1907. Shortly afterward, however, an embarrassingly elementary error was discovered nearly at the beginning of the paper, invalidating his work.

Others weren't even interested in trying. David Hilbert, the German mathematician, was asked why he didn't try to prove Fermat's last theorem, and his answer was, "Before beginning, I should need to put in three years of intensive study, and I haven't that much time to squander on a probable failure." Gauss likewise was uninterested, although for a different reason. According to Paulo Ribenboim's book *Thirteen Lectures on Fermat's Last Theorem,* he wrote in a letter to a colleague, "I am very much obliged for your news concerning the Paris prize. But I confess that Fermat's theorem as an isolated proposition has very little interest for me, because I could easily lay down a multitude of such propositions, which one could neither prove nor dispose of."

[*Unlearned Attempts at Proof*]

After this increased attention, the German Academy of Sciences in Göttingen offered an astonishing prize of 100,000 marks in 1908, actually willed by a German mathematics professor named Paul Wolfskehl. Despondent over a lost love and—just like in the movies—not being able to find a proof of Fermat's last theorem, he decided to commit suicide and made plans for his final act. Philip Davis and William Chinn described the tense hours that followed in their book *3.1416 and All That*. "A few hours remained until the appointed time. He went into his library, wondering what to do. He took down some mathematical pamphlets from the shelf and fingered them idly. By pure chance, he opened one of them. It was Kummer's work on Fermat's last theorem. As he read the article, he thought he spotted an error in Kummer's work. One hour passed, two, three hours, while Wolfskehl checked the mathematics. Finally, he was forced to admit that Kummer's argument was completely sound!" But by then, the appointed hour had passed, and Wolfskehl abandoned his plan.

In order to win the prize willed by Wolfskehl (who later died of natural causes), the proof was required to be published and to be judged correct by the German Academy of Sciences no sooner than two years later. The prize is still standing, although much reduced in value. (The soaring inflation of the Weimar period in the 1920s took its toll, and the prize has now been reduced to 7,500 marks, about $4,400 these days.) Nevertheless, the German prize offer "attracted the world's cranks," as *The New York Times* put it on June 24, 1993, quoting Edwards.

When the Germans said the proof had to be published, "the cranks began publishing their solutions in the vanity press," he said, yielding thousands of booklets. The Germans told him they would even award the prize for a proof that the theorem was not true, Edwards added, saying that they "would be so overjoyed that they wouldn't have to read through these submissions."

[*Who Is a Crank?*]

I understand the sentiment expressed by Edwards and indeed by every credentialed person I have ever heard comment on the subject. And I may be virtually alone in believing that amateur efforts are worthwhile, even though there have been an incredible number of them in the case of F.L.T. But here is my reasoning:

Fermat not only had a habit of scribbling little notes in the margins of books, he also casually mentioned great discoveries over dinners with friends and didn't bother publishing them at all. (His notes—not his formal papers—were published posthumously.) For that reason (and another), he didn't get credit for his independent discovery of analytic geometry, which went entirely to the French philosopher and mathematician René Descartes instead, even though Descartes's analysis was two-dimensional and Fermat's was three-dimensional.

Odd, isn't it? Not really. The "other reason" mentioned in the previous paragraph is that Pierre de Fermat was himself an "amateur." That is, he was never a professional mathematician, which is the main reason he didn't publish papers in the normal fashion. Instead, he studied law and was graduated from the University of Orléans in 1631, later becoming a judge for the parliament in Toulouse. In his book called *Number Theory*, André Weil tells us that "the closest [Fermat] ever came to personal contact with a mathematician (apart from a [possible] visit from [the mathematician Jean de] Beaugrand was [a] three days' meeting with [Father] Mersenne . . . ," who was not a professional mathematician himself. In fact, Fermat's participation in the mathematical community was entirely through private correspondence to noted mathematicians.

Publishing papers was not easy, in any case, as André Weil relates in the same book:

> Actually, in those days, it was not quite a simple matter for a mathematician to send a work to the press. For the printer to do a tolerable job, he had to be closely supervised by the author, or by someone familiar with the author's style and notation; but that was not all. Only too often, once the book had come out, did it become the butt of acrimonious controversies to which there was no end. . . . At the same time, it is clear that he [Fermat] always experienced unusual difficulties in writing up his proofs for publication; this awkwardness verged on paralysis when number theory was concerned, since there were no models there, ancient or modern, for him to follow.

Regardless, this is why I feel that no "amateur" who ever worked on F.L.T. should be called a crank for that reason alone. Not unless we apply that term to Fermat himself. (Or Germain, for that matter.)

Whether we speak of the arts or of the sciences, closing a field of thought to its credentialed followers exclusively can only hinder that field's progress—at least in part because those very credentials so often are channelers of thought. In the past, the problem was that generations of scientists (and artists) were taught certain precepts early in their careers and then adhered to them, right or wrong, producing many didactic professionals who held too dearly to the familiar. But in more modern times, we may be having a philosophical pole-shift and finding that incorrect *new* precepts are taking hold in a similar way.

Has impatience and frustration with the assumed limits of human capacity become so great that intellectual energy now expresses itself in an excess of creativity? Have past generations of scientists, protesting farfetched new ideas to their dying day, finally vanished into the past, leaving future generations to be raised with these same new ideas, effectively turning fantasy into fact, at least for the time being? It may be.

So an attorney was the founder of modern number theory. Why may an accountant not prove his point? Or an insurance salesman? (Other than the obvious fact that their actual chances of success are virtually nil.) These people aren't cranks. (Which brings another point to mind. If, in a letter to Hilbert, an amateur mathematician had written like Eddington did when he described his rotating bubble, would his letter have been sent sailing into the "round file"?)

But many amateur efforts, which clearly include some desperate and/or unbalanced individuals, apparently have been more than amusing, as evidenced in the following letter written in 1974 by F. Schlichting of the Mathematics Institute of the University of Göttingen to Paulo Ribenboim and quoted in his book *Thirteen Lectures on Fermat's Last Theorem*.

> Please excuse the delay in answering your letter. . . . There is no count of the total number of "solutions" submitted so far. In the first year (1907–1908) 621 solutions were registered in the files of the Akademie, and today they have stored about three meters of correspondence concerning the Fermat problem. In recent decades, it was handled in the following way: the secretary of the Akademie divides the arriving manuscripts into (1) complete nonsense, which is sent back immediately, and into (2) material which looks like mathematics. The second part is

given to the mathematical department and there, the work of reading, finding mistakes, and answering is delegated to one of the scientific assistants . . . at the moment, I am the victim. There are about three to four letters to answer per month, and there is a lot of funny and curious material arriving, e.g., like the one sending the first half of his solution and promising the second if we would pay 1000 DM in advance; or another one, who promsied me ten percent of his profits from publications, radio, and TV interviews after he got famous, if only I would support him now; if not, he threatened to send it to a Russian mathematics department to deprive us of the glory of discovering him. From time to time, someone appears in Göttingen and insists on personal discussion. . . .

As for Fermat himself, as André Weil tells us in his book *Number Theory*, "His proofs have almost totally vanished. Writing them up, at a time when algebraic notation was still clumsy in the extreme, and models were altogether absent, would have cost a tremendous effort, and the complete lack of interest on the part of his contemporaries must have been depressing."

Nevertheless, the tide of interest in F.L.T. seemed endless. The problem even turned up in two works of fiction—*Murder by Mathematics*, a mystery novel by Hector Hawton, and *The Devil and Samuel Flagg*, a short story by Arthur Porges. In that story, according to the January 1989 issue of *Discover*, "Simon, the protagonist . . . has managed to persuade Satan to engage in a battle of wits. Simon gets to pose a single question, and the devil has twenty-four hours to come up with the correct answer, for which he'll win the man's soul. Should the devil fail, however, he must provide Simon with lifelong health, happiness, and money." Here's how the conversation went:

"All right," said Simon. He took a deep breath. "My question is this: Is Fermat's Last Theorem correct?"

The devil gulped. For the first time his air of assurance
weakened. "Whose last what?" he asked in a hollow voice.

"Fermat's proposition will 'probably never be proved or
disproved,'" noted Philip Morrison in a review in *Scientific
American* in April 1985.

A few years later, in January 1992, Morrison wrote in another review in the same publication: "[the book] closes in
1988, when one among many entries reports the claim that
the missing proof of Fermat's last teasing theorem had after
three centuries been found! (. . . but flaws were soon found
in the new 'proof'.)" In its edition of April 14, 1988, *New Scientist* quoted Enrico Bombieri of the Institute for Advanced
Study in Princeton. " 'From a mathematical point of view,
[the flawed proof] is not useless,' adding that, there are 'very
interesting ideas involved.' " Gerd Faltings, also of Princeton
University, added, " 'We need a new idea.' " So when we
wonder just how famous is F.L.T., it's amusing to note that
Yoichi Miyaoka, the Japanese mathematician just mentioned,
an expert in differential geometry at Tokyo Metropolitan University, actually made it into the record books with an *errone-
ous* proof.

In his article entitled "Number Theory as Gadfly," Barry
Mazur notes that "Despite the fact that [Fermat's last theo-
rem] resists solution, it has inspired a prodigious amount of
first-rate mathematics."

[*Wiles's Proof in Not-Very-Plain English*]

There is no way to state this proof in plain English, but the
following paragraphs are an effort toward that end.

Equations of the form $x^n + y^n = z^n$, called Diophantine
equations when n is an integer, can be translated to describe a
certain set of elliptic curves. These curves represent the sur-
face of a torus, an object shaped like a smooth doughnut.

Taniyama suggested that for that certain set of elliptic

curves in Euclidean geometry (where parallel lines never meet, even if infinitely extended), there are corresponding structures in the hyperbolic (non-Euclidean) plane (where parallel lines can both converge and diverge).

Frey suggested a connection between that certain set of elliptic curves and Fermat's last theorem, namely that if there were solutions in violation of the theorem, they would generate a subset of "semistable" elliptic curves, curves that could *not* be represented in the hyperbolic plane.

Wiles accepted Ribet's proof of Frey and reasoned that if he (Wiles) could prove Taniyama, at least for the "Fermat subset" of semistable elliptic curves (if not for that larger certain set of elliptic curves), no solutions to Fermat's last theorem could exist, thereby implying a proof of F.L.T.

Wiles then developed an unconventional method of counting both the Euclidean semistable elliptic curves and their hyperbolic (non-Euclidean) counterparts in such a way as to demonstrate a one-to-one correspondence between the two groups. In this way, he claims to have proved Taniyama for the "Fermat subset" of semistable elliptic curves.

[*Unanswered Questions*]

Other questions come to mind. If Fermat himself proved that his last theorem was true when the exponent was 4, then why didn't he offer a general proof? The notes in the margin are supposed to have been written in 1637, but Fermat didn't die until 1665, almost thirty years later. (It sounds as if he had plenty of time.) And how do we know the date of 1637 in the first place? Did Fermat date his hastily written little note? Ah, but the answers to those questions, as the answers to so many other questions, are lost in time.

3

Proofs and Puzzles to Ponder

[*What's a Theorem?*]

WITHOUT STOPPING TO think about it, try to guess how many formal theorems are published yearly in mathematical journals. A hundred? A thousand? Ten thousand?! Stanislaw Ulam disabuses us of our innocence in his autobiography *Adventures of a Mathematician*.

> At a talk which I gave at a celebration of the twenty-fifth anniversary of the construction of von Neumann's computer in Princeton a few years ago, I suddenly started estimating silently in my mind how many theorems are published yearly in mathematical journals. I made a quick mental calculation and came to a number like one hundred thousand theorems per year. I mentioned this, and my audience gasped. The next day, two of the younger mathematicians in the audience came to tell me that, impressed by this enormous figure, they undertook a more systematic

and detailed search in the Institute library. By multiplying the number of journals by the number of yearly issues, by the number of papers per issue, and the average number of theorems per paper, their estimate came to nearly two hundred thousand theorems a year. If the number of theorems is larger than one can possibly survey, who can be trusted to judge what is 'important'? One cannot have survival of the fittest if there is no interaction.

Two hundred thousand theorems a year! On July 3, 1993 — shortly after Wiles's proposed proof—*Science News* noted that "Fermat's last theorem may finally live up to its common designation as a theorem." But just what *is* a theorem? We'll start at the beginning.

What's an Axiom? What's a Postulate? The mathematical distinction between these two words is insignificant, and most mathematicians use them almost interchangeably. Often, they'll use the term "axiom" when they personally agree with the principle in question, and they'll use the term "postulate" when they feel less charitable. Either way, an axiom or postulate is an initial assumption that is accepted as true without proof, and from which conjectures are derived.

What's a Theorem? A theorem is a conjecture that has been proved in a formal argument using axioms and other theorems. (Without a proof, F.L.T. more properly should have been called "Fermat's Last Conjecture,") which explains the remark in *Science News*. There are different varieties of theorems. Some are more substantial than others, some are favored more or less, and some have certain characteristics.

A "uniqueness theorem," for example, shows that one — and only one—mathematical entity satisfies a particular condition. Often, this is proved by assuming more than one solution and showing that this leads to a contradiction.

An "existence theorem" shows only that *some* mathematical entity satisfies a particular condition. The proof may be indirect, and it need not actually produce the entity, but it must show that there is such a solution.

[*What's a Proof?*]

Some mathematicians object philosophically to existence theorems, or "existence proofs," as they're also known, because they're "nonconstructive proofs." They demand a "constructive proof" instead. A constructive proof not only shows that one (or more) mathematical entity satisfies a particular condition, it also actually produces it (or them). Not surprisingly, constructive proofs are often much longer, more complex, and harder to formulate than existence proofs. There are several varieties of these, also. Perhaps the most popular compromise is known as the intuitionist view. It demands a certain limited kind of constructive proof, but "proof by contradiction" is strictly not permitted.

A "proof by contradiction" (also called "proof by double negation") is a weaker proof and is rejected entirely by some. It proves a theorem by assuming that the conjecture is false and finding that this leads to a contradiction. This is also known as *"reductio ad absurdum."* When a contradiction is found, the conjecture is then proved to be true.

"Proof by double negation" is fraught with dangers. For example, let's say the human animal can see only in black and white. (To make this point, let's also assume that nothing is gray.) We want to prove that an object is white, so we manage to prove that it isn't black. By double negation, the object is proved to be white. But considering all the colors of the rainbow, this shouldn't be a valid proof. Maybe the object is red. But how would we know it?

Reductio ad absurdum is an "indirect proof." Its opposite is known as a "direct proof," which satisfies the most stringent requirements. Proof devolves by a step-by-step process based on axioms or other theorems, especially theorems also proved by direct means.

An effort to maintain high standards is sometimes not possible, often not practical, and too often not profitable. Philip Davis and Reuben Hersh make the following comments in their book *The Mathematical Experience*.

> The actual situation is this. On the one side, we have real mathematics, with proofs which are established by "consensus of the qualified." A real proof is not checkable by a machine, or even by any mathematician not privy to the gestalt, the mode of thought of the particular field of mathematics in which the proof is located. Even to the 'qualified reader,' there are normally differences of opinion as to whether a real proof (i.e., one that is actually spoken or written down) is complete or correct. These doubts are resolved by communication and explanation, never by transcribing the proof into first-order predicate calculus. Once a proof is "accepted," the results of the proof are regarded as true (with very high probability). It may take generations to detect an error in a proof.

Not very reassuring, is it? Another common misperception is to be more impressed by longer proofs than by shorter ones. This may come from mistakenly equating length with difficulty instead of with technicality. (The analogy that comes to mind is a Rube Goldberg–like contraption that consists of a wildly complex machine, a whole kitchenful of apparatus to, say, make a pot of coffee. If that's the only way it can be made, it makes sense to be pleased with it, but it doesn't make sense to be impressed by its sheer size alone.) And with inductive proofs, the longer ones—especially those that run to hundreds of pages—are actually *weaker* than shorter ones.

The following are some informal examples of "proofs."

[*Faulty* ''*Proofs*'']

Martin Gardner puts forth the following "proof" of the parallel postulate in *Wheels, Life and Other Mathematical Amusements*.

> A simple proof of the parallel postulate uses the diagram shown [below]. AB is the given line and C the outside point. From C drop a perpendicular to AB. It can be shown, without invoking the parallel postulate, that only one such perpendicular can be drawn. Through C draw EF perpendicular to CD. Again, the parallel postulate is not needed to prove that this too is a unique line. Lines EF and AB are parallel. Once more, the theorem that two lines, each perpendicular to the same line, are parallel is a theorem that can be established without the parallel postulate, although the proof does require other Euclidean assumptions (such as the one that straight lines are infinite in length) that do not hold in elliptic non-Euclidean geometry. Elliptic geometry does not contain parallel lines, but given Euclid's other assumptions one can assume that parallel lines do exist.
>
> We have apparently now proved the parallel postulate. Or have we?

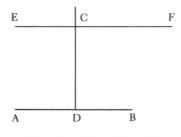

A "proof" of the parallel postulate

No, we haven't. This is Gardner's explanation:

> The proof is valid in showing that one line can be constructed through C that is parallel to AB, but it fails to prove that there is only one such parallel. There are many other methods of constructing a parallel line through C; the proof does not guarantee that all these parallels are the *same* line. Indeed, in hyperbolic non-Euclidean geometry an infinity of such parallels can be drawn through C, a possibility that can be excluded only by adopting Euclid's fifth postulate or one equivalent to it. Elliptic non-Euclidean geometry, in which *no* parallel can be drawn through C, is made possible by discarding, along with the fifth postulate, certain other Euclidean assumptions.

This next "proof" that $2 = 1$ is also faulty, of course.

Let's start with the simple equation:	$A = B$
Multiply both sides by A:	$A^2 = AB$
Subtract B^2 from both sides:	$A^2 - B^2 = AB - B^2.$
Factor both sides:	$(A + B)(A - B) = B(A - B).$
Divide both sides by $(A - B)$:	$A + B = B$
As we know that $A = B$, this means:	$B + B = B$
Add the 2 B's:	$2B = B$
Divide by B:	$2 = 1$

What's interesting about this faulty proof is that it illustrates the problem encountered with more subtle errors. Notice that at one point, we divided both sides of the equation by $(A - B)$, which is usually perfectly permissible. However, because A and B were equal in this case, A minus B equals zero, and dividing by zero is not a permitted mathematical operation.

It seems obvious in retrospect, but consider the problems with detecting truly subtle errors that occur in long, intricate calculations that describe concepts completely theoretical in nature. In the above case, the error would be difficult to detect if we didn't know that A = B. How would we find the error otherwise?

[*Impractical Proofs*]

There are also plenty of ingenious concepts that seem to work perfectly well in theory, but are worth little in practice. Here's an example:

You'd like to move to a desert island and take your set of encyclopaedia along with you, but your parachute has no place for baggage. How can you carry the entire set in your pocket?

Let's say there are fewer than a thousand different letters and numerals (and so forth) in the encyclopaedia. You simply assign them each a code number from 001 through 999, assigning a number to the space between words, too, and any other symbols that are necessary. (For example, the word "a" could be encoded as 001, the word "at" could be encoded as 001020, and the word "ate" could be encoded as 001020005.) Next, you write down all these code numbers, making one immense number, and then place a decimal point in front of it to make the number a decimal fraction. (The word "ate" would become .001020005.) Now take a little ten-centimeter ruler and make a mark on it, dividing it into two parts, segment A and segment B, such that the fraction A/B is equal to the decimal fraction. (Such as A/B = 2.4/7.6 = .31578947368 . . .) When you get to the desert island, you look (carefully!) at the lengths of A and B, then compute the fraction A/B, and the code for the entire encyclopaedia will be in front of you.

(Then again, consider how a few years ago, it would have

seemed ridiculous to find all twelve volumes of the original Oxford English Dictionary, an immense amount of material, encoded onto a single compact disc, a copy of which I have here at my desk.)

[*Branching Proofs*]

The following problem sounds easy at first, but quickly becomes ponderous due to the branching nature of the solution. It gives a glimpse into the many paths that must all be followed in order to pin down a proof, and the sheer length of it is remarkable. (However, this is not an example of a lengthy proof, which follows point after point in a much more serial fashion.) Here's the problem:

> You have twelve balls that appear identical, except that one is slightly heavier or lighter than the others. Using only a balance scale, you have three weighings to discover which is the "odd ball" and whether it's lighter or heavier.

And here's the answer:

For the first weighing, put two groups of four balls in each pan of the scale. Set the remaining four aside. If the scale balances, continue with (A) *If the First Weighing Balances*. If it doesn't, skip directly to (B) *If the First Weighing Doesn't Balance*.

(A) If the First Weighing Balances:

If the first weighing balances, you know all of those eight balls are normal. For the second weighing, leave just three of these normal balls on one side of the scale, put the other five normal balls away, and put three unknown balls (from the four you set aside) on the other side of the scale. If it balances again, read Number 1 below. If it doesn't balance, read Number 2.

1. *If the second weighing balances*. Now you know the last

unknown ball (from the original four you set aside) is the odd ball. For the third weighing, leave just one normal ball on one side of the scale, take the additional five normal balls off the scale and put them away, and put the last unweighed odd ball on the other side. It either will rise or fall, telling you whether it's light or heavy. That's the answer for this sequence.

2. *If the second weighing doesn't balance.* But now you know whether the odd ball is light or heavy, because the group (on the scale) where it's hidden either rose (if it's light) or fell (if it's heavy). For the third weighing, first put away all the normal balls, then take that "odd group" of three already on the scale and separate them all, putting one on each side of the scale and leaving one aside.

Maybe the scale will balance again. If it does, you'll know that the last ball you set aside is the odd ball, and because you already know whether it's light or heavy, that's the answer for this sequence of events.

Or maybe the scale will tilt. If it does, this will show you which is the odd ball because you already know that it's either light or heavy. If it's light, the odd ball is the one that rises, and if it's heavy, the odd ball is the one that falls. And that's the answer for this sequence of events.

(B) If the First Weighing Doesn't Balance

Now things are going to get a little more complicated to describe, so let's number the balls. Numbers 1, 2, 3, and 4 will be the ones on the lower side of the scale, and Numbers 5, 6, 7, and 8 will be the ones on the higher side. The odd ball must be among these eight. Numbers 9, 10, 11, and 12 will be the ones that have been set aside, so these four must all be normal. That much we know.

For the second weighing, leave Number 1 on the same side (previously lower) of the scale and put Numbers 2, 3, and 4 aside. Instead of those three, substitute Numbers 6,

7, and 8 (three of the ones that were previously on the higher side of the scale). Leave Number 5 on the same side of the scale (previously higher), but add three normal balls (from the original group of four you set aside; we won't need the last normal one any more). At this point, you have Numbers 1, 6, 7, and 8 on what was previously the lower side, and you have Numbers 5, 9, 10, and 11 on what was previously the higher side. Number 12 can be tossed into the closet.

The scale must now do one of three things: It will balance, it will tip in the opposite direction as it did before, or it will tip in the same direction it did before.

1. *If the second weighing balances.* Now you know that Numbers 2, 3, and 4 (newly set aside) contain a heavy odd ball. Why? Because when they were on the scale in the first weighing, they pulled it down against others now proven to be normal. For the third weighing, put away all the normal balls, then take that odd group of three currently set aside and separate them all, putting one on each side of the scale and leaving one aside again.

Maybe the scale will balance again. If it does, you'll know that the last ball you set aside is the odd ball, and because you already know it's heavy, that's the answer for this sequence of events.

Or maybe the scale will tilt. If it does, this will show you which is the odd ball, because you already know that it's heavy. And that's the answer for this sequence of events.

2. *If the second weighing tips in the opposite direction.* Now you know that Numbers 6, 7, and 8 (which changed positions on the scale) contain a light odd ball. Why? Because Numbers 1 and 5 didn't change positions, and the others on the scale are known normal. For the third weighing, put away all the normal balls, then take that odd group of three currently on the scale and separate them all, putting one on each side of the scale and leaving one aside.

Maybe the scale will balance again. If it does, you'll know that the last ball you set aside is the odd ball, and because you already know that it's light, that's the answer for this sequence of events.

Or maybe the scale will tilt. If it does, this will show you which is the odd ball because you already know that it's light. And that's the answer for this sequence of events.

3. *If the second weighing tips in the same direction.* Now you know that Numbers 1 or 5 is the odd ball. Why? Because they're the only ones on the scale that didn't change position. If the odd ball were among Numbers 6, 7, and 8, the scale would have reversed, and the others on the scale are known to be normal. In addition, you know that either Number 1 is heavy or Number 5 is light because that's the direction in which the scale tips.

For the third weighing, clear everything from the scale. Then put one normal ball on one side and Number 1 on the other.

Maybe the scale will balance again. If it does, you'll know that Number 5 is the odd ball, and because you already know that it's light, that's the answer for this sequence of events.

Or maybe the scale will tilt downward toward Number 1. If it does, you'll know that Number 1 is the odd ball, and because you already know that it's heavy, that's the answer for this sequence.

[*Philosophical Proofs*]

Let's think about wheels again, but this time from the perspective of confirmation theory. Suppose I see half a dozen wheels that are round, and I conjecture that "All wheels are round." Then I see half a dozen more, all of which are round, too. The conjecture is weakly confirmed. And if I see thousands and thousands of wheels, all of them round, the conjecture is more strongly confirmed. But it's not fully confirmed. Maybe there are some wheels that aren't round, but I just haven't seen any.

What about a paper clip? Can this be a confirming instance of my conjecture? At first, a paper clip seems to have nothing to do with it, but consider this. Here's another way of stating the wheels conjecture that is logically equivalent. "All objects that aren't round are not wheels." Then I find a paper clip, which isn't round and isn't a wheel, either. Clearly, this is a confirming instance of the logically equivalent conjecture of "All objects that aren't round are not wheels." For that reason, it must also confirm the original conjecture of "All wheels are round." Apparently, it's not going to be hard to prove my original conjecture because I simply can look around my office and find dozens of objects that aren't round that are not wheels. (Thousands, actually. I have whole boxes full of paper clips nearby.)

But we shouldn't stop here. Let's say that we decide that a paper clip is a confirming instance of the original conjecture, but only to a minuscule degree. (This is what the original theorist, the German-born American philosopher Carl Gustav Hempel, believed.) Let's even say that some confirming instances (like finding a round wheel) count for more than others (like finding a paper clip). Then let's try a new conjecture entirely: "All wheels are square." Here's another way of stating the new conjecture that is logically equivalent. "All objects that aren't square are not wheels." Then I find a paper clip, which isn't square and isn't a wheel, either. Clearly, this is a confirming instance of the logically equivalent conjecture of "All objects that aren't square are not wheels." For that reason, it must also confirm the original conjecture of "All wheels are square."

But this means that finding a paper clip confirms that "All wheels are round," *and* "All wheels are square," simultaneously! How can that be? (Now you know why these issues are not black and white. Or would that be "white and black"?)

[*What Is Mathematical Truth?*]

That's a little trickier to define. In 1900, David Hilbert, one of the most famous and respected mathematicians of the day, offered a list of unsolved problems to his colleagues at the International Congress of Mathematicians in Paris. The twenty-third was a challenge to discover a precise method for determining the truth (or falseness) of any given statement in formal logic. The issue was settled, in a fashion, some thirty-six years later, but surely not in the way Hilbert originally had anticipated. After all, if such a method for determining the truth were found, it would seem to reduce the science of mathematics to mere mechanical computation. No wonder few mathematicians wished to champion that cause.

Two developments finally closed Hilbert's line of inquiry. First, Bertrand Russell, the British philosopher and mathematician, discovered a paradox in the elementary theory of sets in 1901, just as Gottlob Frege, the German philosopher and mathematician, was about to publish his *Fundamental Laws of Arithmetic,* which not only depended upon the theory of sets, but would have furthered Hilbert's investigation. According to the May 1984 issue of *Scientific American,* "Frege ended the volume with a dispirited note: 'A scientist can hardly meet with anything more undesirable than to have the foundation give way just as the work is finished. I was put in this position by a letter from Mr. Bertrand Russell when the work was nearly through the press.' "

Then Kurt Gödel, of the Institute for Advanced Study in Princeton, New Jersey, dealt the final blow. In the May 1984 issue of *Scientific American,* it was noted that "he proved that any consistent system of formal logic powerful enough to formulate statements in the theory of numbers must include true statements that cannot be proved. Because consistent axiomatic systems such as the one devised by Russell and [Alfred North] Whitehead [the English mathematician and philosopher] cannot encompass all the true statements in the subject matter they seek to formalize, such systems are said to be incomplete." One wonders whether Gödel's logic could be applied to his own argument, like a dog chasing its tail, thus "proving" itself invalid.

So much for the truth. The search for it has never recovered from this defeat. Raymond Wilder addressed the subject in his book *Evolution of Mathematical Concepts.*

> As a peculiarly Greek mode of thought, logic lay at the very heart of the axiomatic method. And it need hardly be pointed out what the ascendance of logic in mathematics has meant to mathematics. Moreover, its importance in methods of proof is so great that ultimately some came to insist that mathematics is really an *extension* of logic, maintaining that the *essence* of mathematics is logical deduction. . . . However, it seems safe to say that this point of view has not many supporters today.

> . . . [After the advent of non-Euclidean geometries] The dual character of mathematics was retained; mathematics was still an instrument for scientific investigation, but on the conceptual side, it now achieved a freedom that it did not know before. This was freedom accompanied by a conviction that it was no longer restrained by either an ideal or an external world, but that it could create mathematical concepts without the restrictions that might be imposed by either the world of experience or an ideal world of "truth" to whose nature one was committed to limit his discoveries. This feeling of freedom was not entirely justified, but for the time being, it was a grand tonic.

[4]

Mathematical Mysteries, Solved and Unsolved

[The Parallel Postulate]

THE GRANDEST OF passions are evoked by mathematical mysteries, as evident in the following quote from a letter in Philip Davis and Reuben Hersh's book *The Mathematical Experience*. In it, the Hungarian mathematician Farkas (Wolfgang) Bolyai was advising his son János Bolyai, also a mathematician and who would one day become a very famous one indeed, not to spend any more time pursuing his course on the notorious "parallel postulate (Euclid's fifth axiom)."

> For God's sake, please give it up. Fear it no less than the sensual passions because it, too, may take up all your time and deprive you of your health, peace of mind, and happiness in life.

H. Meschkowski's book *Evolution of Mathematical Thought* quotes additional tortured pleas from father to son.

You must not attempt this approach to parallels. I know this way to its very end. I have traversed this bottomless night, which extinguished all light and joy of my life. I entreat you, leave the science of parallels alone . . . I turned back when I saw that no man can reach the bottom of this night. I turned back unconsoled, pitying myself and all mankind. . . .

. . . It seems to me that I have been in these regions; that I have travelled past all reefs of this infernal Dead Sea and have always come back with broken mast and torn sail. The ruin of my disposition and my fall date back to this time. I thoughtlessly risked my life and happiness—*aut Caesar aut nihil.*

Oh, the *pain* of it all! In Fyodor Dostoyevsky's novel *The Brothers Karamazov*, Ivan (speaking to his brother) is very firm about his position.

But there's this that has to be said: if God really exists and he really has created the world, then, as we all know, he created it in accordance with the Euclidean Geometry, and he created the human mind with the conception of only three dimensions of space. And yet there have been and there still are mathematicians and philosophers, some of them indeed men of extraordinary genius, who doubt whether the whole universe, or, to put it more widely, all existence, was created only according to Euclidean geometry, and they even dare to dream that two parallel lines which, according to Euclid, can never meet on earth, may meet somewhere in infinity. I, my dear chap, have come to the conclusion that if I can't understand even that, then how can I be expected to understand about God? . . . Let . . . the parallel lines meet, and let me see them meet, myself—I shall see, and I shall say that they have met, but I still won't accept it.

The parallel postulate has probably caused mathematicians more grief than anything in their history. Indeed, it was the failure of Wolfgang Bolyai (quoted at the beginning of this section) or his son to prove the parallel postulate that led János to independently develop hyperbolic geometry (along with the German mathematician Carl Friedrich Gauss and the Russian mathematician Nikolai Lobachevsky), which allows an infinite number of parallels. (The German mathematician Georg Friedrich Bernhard Riemann developed elliptic geometry, which allows none.)

In his book *Wheels, Life and Other Mathematical Amusements*, Martin Gardner adds, "The eighteenth-century French geometer Joseph Louis Lagrange was convinced that he had produced such a proof. . . . In the middle of the first paragraph of a lecture to the French Academy on his discovery, however, he suddenly said, *'Il faut que j'y songe encore'* (I shall have to think it over again), put his papers in his pocket, and abruptly left the hall."

This is not an idle pursuit, by any means. Because the general theory of relativity requires a non-Euclidean universe, one could shatter Einstein's work by exposing a contradiction in non-Euclidean geometry, which would be accomplished by a proof of the parallel postulate.

[*A r i s t o t l e ' s   W h e e l*]

Another mind-bending problem is the Greek philosopher Aristotle's wheel paradox. Modern mathematicians are not troubled by this one, but nearly everyone else is, though.[1] It dates back to Aristotle's time, when it was first mentioned in the Greek *Mechanica*, later attributed to him. This paradox was

---

[1] The resolution of the paradox lies with the concept of densities of infinities and the correspondingly unique qualities of transfinite numbers; the number of points on any segment of a curve is described by the second of the transfinite numbers, known as aleph-one.

studied by numerous prominent mathematicians, including Galileo in the sixteenth century, Descartes in the seventeenth century, and Fermat himself.

Aristotle's wheel

As the larger wheel rolls along from point A to point B, the smaller wheel rolls along a parallel line from point C to point D. Let's assume that the larger wheel rolls along from point A to point B without slipping. (We make this assumption to define one of the two related problems. As Martin Gardner puts it in *Wheels, Life and Other Mathematical Amusements*, "If the two lines are actual tracks, the double wheel obviously cannot roll smoothly along both. It either rolls on the upper track while the large wheel continuously slides backward on the lower track, or it rolls on the lower track while the small wheel slips forward on the upper track. This is not, however, the heart of the paradox.")

At any given moment in time, there is a singular point on the perimeter of the large wheel touching the line AB, and at that same moment in time, there is a singular point on the perimeter of the small wheel touching the line CD. That is, every point on the large wheel can be associated with every point on the small wheel and vice versa. This appears to prove that the two perimeters are the same length. Clearly, that isn't true, and this is a good example of how a seemingly correct proof can be very much in error. In the above case, the fact that the proof is in error is obvious visually, if not intellectually. But as the proof becomes longer and more complicated, errors become progressively more difficult to locate and may even be impossible to detect, especially in theoretical territory.

[*Zeno's Achilles*]

The Greek philosopher Zeno loved to confound his contemporaries by posing seemingly unsolvable paradoxes, which he did in abundance. However, the one most often cited is the paradox known as "Achilles." Suppose that Achilles and a tortoise agree to run a race with each other. Achilles can run ten times as fast as the tortoise, but he gives the tortoise a 100-meter start. As Zeno put it, Achilles runs 100 meters and arrives at the place where the tortoise started. Meanwhile, the tortoise has run 10 meters and is therefore that far ahead of Achilles. While Achilles runs that 10 meters, arriving at the place where the tortoise was, the tortoise has run one meter and is that far ahead of Achilles. And while Achilles runs that one meter, arriving again at the place where the tortoise was, the tortoise has run one-tenth of a meter more and is still that far ahead of Archilles. (You know where this leading, don't you?)

Let's stop here. This reasoning appears to prove that Achilles will never catch up with the tortoise — that the tortoise will always be ahead by some distance even if that distance is infinitesimally small. But again, that clearly isn't true, and this is another good example of a faulty proof. The fact that the proof is in error is obvious; we know that Achilles will eventually pass up the tortoise. But how? And how would we possibly see this sort of error if it dealt with concepts that don't exist in the real world?

In this case, the key is the concept of time. However fast or slow they run, time will pass. Assign any time (see below for an example), and you'll discover that Achilles overtakes the tortoise at 111.111 (and so on) meters. The repeating decimals may seem to preclude a final answer, but the understanding that they simply represent the fraction ⅑ makes it easier to accept. You can easily envision ⅑ of a pie; that same slice is represented by the number 0.111 (and so on).

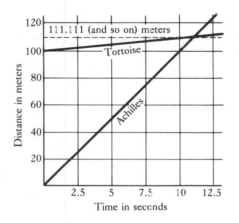

Another of Zeno's paradoxes is called "The Dichotomy," which can be illustrated with a runner. She starts running at a steady pace, but before she gets to the finish line, she must arrive at the halfway point. Then she must arrive at the ¾ mark, which is half of the remaining distance. Then she must arrive at the ⅞ mark, half of the remaining distance. And then she must arrive at the ¹⁵⁄₁₆ mark, half of the remaining distance again. As she always has another halfway point ahead of her, how will she ever get to the finish line?

This reasoning appears to prove that she never will, but we know that isn't true. Again, the key is time. Let's suppose the runner takes a minute to run each half-distance; she gets closer and closer to the finish line, but she never arrives there, as in (A). This reasoning is incorrect. The runner cannot take a minute to run each half-distance because she would be slowing down if she did; instead, she runs at a steady pace (each half-distance is run in half the time it took to run the previous one) as in (B). If she takes a minute to run the first half of the race, she'll take a minute to run the second half of the race.

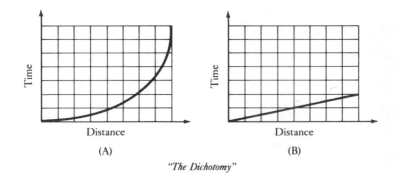

(A)   (B)

"The Dichotomy"

However, even these time-related proofs are not as satisfying as they could be. Zeno could argue that just as there is always another halfway point in distance, there is always another halfway point in time, and that's certainly true. In fact, as recently as 1914, Bertrand Russell, the British philosopher and mathematician, argued very effectively in his book *Our Knowledge of the External World* (and many agree with him) that Zeno's paradoxes were not convincingly answered until Georg Cantor, the German mathematician, revolutionized mathematics with his work on transfinite numbers and set theory. (See previous footnote.)

[*Zeno's Arrow*]

The third of Zeno's paradoxes is known as "The Arrow." Zeno argued that an object, illustrated by an arrow, can either stand still or it can move. If it's standing still, it occupies a space of its own length. Yet this seems to be the case even when the arrow is flying through the air—it always occupies a space of its own length. (Today's camera technology captures this argument on film. If several rapid photographs of a flying arrow are taken with a fast-enough shutter speed, the arrow will appear quite still in each, but it will have moved from point to point to point.) So the paradox is that there are two axioms that are both true but that seem to contradict each other: An arrow in flight always occupies a space of its own length, but an object that always occupies a space of its own length is standing still. How can the arrow be in motion and standing still at the same time?

Again, Cantor comes to the rescue with transfinite numbers and set theory, but in a less satisfying manner. After all, a photograph of a moving arrow and a still arrow (say, suspended by invisible threads) do look exactly the same. In short, the literature on this paradox is voluminous, and it seems there are no easy answers. The jury is still out on this one.

[*Zeno's Stadium*]

The fourth (and least known) of Zeno's paradoxes is called "The Stadium," but it has been handed down to us in fragmentary form and is difficult to describe without a bit of imaginative reconstruction. Suffice it to say that it involves four stationary bodies of equal dimensions, with four moving bodies heading to the right of them and four more moving bodies heading to the left of them, passing each other in the process. This is the most controversial of the paradoxes, and if you're interested in a fuller explanation, you'll find it in Wesley Salmon's book called *Zeno's Paradoxes*, in which he makes this comment:

> Zeno's paradoxes have an onion-like quality; as one peels away outer layers by disposing of the more superficial difficulties, new and more profound problems are revealed. For instance, as we show that it is mathematically consistent to suppose that an infinite series of positive terms has

a finite sum, the problem of the infinity machines arises. When we show how the infinity machines can be handled, the problem of composing the continuum out of unextended elements appears. When charges concerning the consistency of the continuum are met, the problem of identity of structure between the mathematical continuum and the continuum of physical times confronts us. And so on. Will we ever succeed in stripping away all of the layers and providing a complete resolution of all the difficulties that arise out of Zeno's paradoxes? And if we should succeed, what would be left in the center? In a certain sense, nothing, it would seem. We will not find a metaphysical nutmeat . . . or any other fundamental truth about the nature of reality. However, we should not conclude that nothing of value remains. The layers we have peeled away have in them the elements of a nourishing philosophical broth. The analysis itself, dealing in detail with a host of fundamental problems, is richly rewarding in terms of our understanding of space, time, motion, continuity, and infinity. We would be foolish indeed to conclude that the onion was nothing but skin, and to discard the whole thing as worthless.

It would, of course, be rash to conclude that we had actually arrived at a complete resolution of all problems that come out of Zeno's paradoxes. Each age, from Aristotle on down, seems to find in the paradoxes difficulties that are roughly commensurate with the mathematical, logical, and philosophical resources then available. When more powerful tools emerge, philosophers seem willing to acknowledge deeper difficulties that would have proved insurmountable for more primitive methods. We may have resolutions which are appropriate to our present level of understanding, but they may appear quite inadequate when we have advanced further. The paradoxes do, after all, go to the very heart of space, time, and motion, and these are profoundly difficult concepts.

Or, has the onion infinitely many layers? If so, we may be faced with an infinite sequence of tasks that does defy completion in a finite time, for the steps become longer, not shorter, as the difficulties become deeper.

The following classic problems remain unsolved.

[*Kepler's Spheres*]

Johannes Kepler, the German astronomer, asked this question: If we're given more than enough spheres to fill a given volume of space, what would be the most efficient way to pack them into that space? Many of us would guess that the best way would be to pack the bottom layer in rows and then pack the next layer by putting a sphere in each of the little spaces made by every three or four spheres beneath it, nesting them back and forth like this row after row. But is that really the best way to fit in the greatest number?

[*Goldbach's Conjecture*]

Christian Goldbach, the German-Russian mathematician, suggested that every even number greater than 2 was the sum of two prime numbers, for example, $4 = 2 + 2; 6 = 3 + 3; 8 = 3 + 5$, and so forth. No one has ever been able to prove that Goldbach was correct, nor has anyone ever been able to find an exception to his rule.

[*The Twin Prime Conjecture*]

Is there an infinite number of "twin primes" (pairs of prime numbers with a difference of 2)? These would be pairs of primes such that one prime $+ 2 =$ another prime, for example, 3 (prime) + 2 = 5 (also prime); 5 (prime) + 2 = 7 (also prime); 11 (prime) + 2 = 13 (also prime).

[*Another Prime Conjecture*]

Is there an infinite number of primes of the form $n^2 + 1$ where n is an integer? Or do they end somewhere?

Although these problems are the most famous, they aren't the last of the classic unsolved problems. In fact, even Fermat has another! (The details are in Chapter Five.)

[5]

What Comes Next

PEOPLE HAVE ALWAYS loved to solve problems, and the people of Fermat's time were no exception. The following volume, written by Jean Leurechon, a French Jesuit writing under a pen name, went through more than thirty editions between 1624 and 1700. This was the title of the first English translation in 1633, just four years before the date Fermat is believed to have made his notorious marginal note.

> *Mathematicall Recreations, or a Collection of Sundrie Problemes, Extracted Out of the Ancient and Moderne Philosophers, as Secrets in Nature, and Experiments in Arithmeticke, Geometrie, Cosmographie, Horologographie, Astronomie, Navigation, Musicke, Opticks, Architecture, Staticke, Machanicks, Chimestrie, Waterworkes, Fireworks, etc., not Vulgarly Made Manifest Until This Time . . . Most of Which Were Written First in Greeke and Latine, Lately Compiled in French, by Henry Van Etten Gent.*

And Now Delivered in the English Tongue with the Examinations, Corrections, and Augmentations.

But non-mathematicians would be mistaken to think that the challenge in proving F.L.T. lay in the Mount Everest sort of life philosophy—one does it "just because it is there." And it's no game or idle pleasure. Wiles's proof may make the work of other number theorists irrelevant and may have many applications in the field of number theory, which is used in the development of computer technology and security codes. In his article entitled "Number Theory as Gadfly," Mazur makes the following comment:

> Despite the fact that [Fermat's last theorem's] truth hasn't a *single* direct application (even within number theory!) it has, nevertheless, an interesting *oblique* contribution to make to number theory; its truth would follow from some of the most vital and central conjectures in the field.

On June 24, 1993, a day after Wiles announced his proof in Cambridge, *The New York Times* quoted Mazur further:

> A lot more is proved than Fermat's last theorem. One could envision a proof of a problem, no matter how celebrated, that had no implications. But this is just the reverse. This is the emergence of a technique that is visibly powerful. It's going to prove a lot more.

For one thing, although much of abstract mathematics seems to be about building castles in the air, that too is not without purpose—although the purpose is not necessarily known ahead of time. Every mathematician hopes that his or her work will find concrete application, if not in modern technologies, then perhaps in modern cosmologies. Einstein's universe is only one example. On June 24, 1993, *Reuter's* reported that Goddard had pointed out that "solving the theorem had no obvious practical application, but commented that the same was said initially about splitting the atom."

For another thing, the concept of isomorphism (or mathematical equivalence) is one of the most fruitful concepts in mathematics. Wiles's proof may have similarly great impact, as many other mathematicians' proofs already have, reverberating far beyond its original locus. Sometimes an intransigent problem will yield to a solution by transposing it into an isomorphic one that already has been solved. One of the best-known examples in modern times is the famous four-color map theorem. When it was proved in 1976, numerous other important (and isomorphic) conjectures were proved true. In other words, this wasn't simply an amusement for cartographers.

The four-color map theorem was posed back in 1850 and attained public recognition well before the turn of the century. The problem is this: How many different colors are necessary to color any map in such a way so that no two adjoining areas are alike? Four colors seemed a likely answer, but are four colors always sufficient? That is, can a map be drawn for which five colors are necessary? It wasn't until 1976 that mathematicians managed to prove that four colors will suffice for all possible configurations, and it took them hundreds of pages to do so. (Wiles's proof, by contrast, was "only" two hundred pages long. Surely, mathematicians must be among the most persevering and best-organized of professionals—at least at the office.)

And F.L.T. itself is too narrow a focus. According to the July 2, 1993, issue of *Science*, "Most number theorists had abandoned [the search for a proof for Fermat's last theorem] as a quixotic quest—. . . one with little payoff beyond the theorem itself. Unlike many other famous unsolved problems in mathematics, Fermat's last theorem has no particularly important consequences." But the magazine added, ". . . while Fermat's last theorem has little practical importance, the Taniyama-Weil conjecture has been of keen interest to number theorists since the mid-1950's, because, if true, it would provide a powerful tool for studying the number-theoretic properties of elliptic curves, which themselves are fundamental in many parts of number theory."

[*Why Wiles's Proof Isn't as Satisfying as It Could Be*]

But as far as Fermat is concerned, Wiles's proof isn't as satisfying as it could be. In addition to the issues raised in Chapter One of this book, other points should be considered. For one thing, mathematicians are reported to have said that the logic of the proof is persuasive because it is built on a carefully developed edifice of mathematics that goes back more than thirty years and is widely accepted. There are two main weaknesses here. First, thirty years is a very short time for a new edifice, and second, how can this logic be widely accepted when almost no one outside of a few specialists understands it? If it is indeed accepted, it must have been accepted on faith.

Also, how do we verify a proof with non-Euclidean geometries, which state that conceptions need not seem possible? Aren't we likely to verify too many proofs this way?

How do we know a contradiction when we see one? And if we believe we have a contradiction, how do we know whether it's a contradiction that proves the theorem or a contradiction in the geometry in which the theorem is based?

How do we know which things *are* impossible and which things merely *seem* impossible, but aren't? For example, how did Eddington know to classify his rotating bubble as possible? How would he decide to rule out other conceptions (which *are* impossible), but accept the rotating bubble (which merely *seems* impossible)? Or is *everything* possible? And if not, how do we tell the difference?

And if conceptions needn't seem possible, proofs are only being held to the standard of following logically from previous proofs, which presumably didn't need to seem possible, either.

There's another philosophical point to consider. In Ribet's paper on the modular representations of Gal ($\bar{Q}/Q$) arising from modular forms (the paper in which he presents proof of Frey, which is the foundation upon which Wiles's proof is built), the third to the last sentence from the end of the paper (published in *Inventiones Mathematicae* in 1990) reads as follows: "The Main Theorem applied inductively to p now eliminates all odd primes from its level." And, of course, Wiles's proof (highlights of which can be found in the Appendix) is also inductive. Both can be considered proofs by double negation. (Mathematical induction is used especially for series sums; it is also used in *reductio ad absurdum*. In these cases, a contradiction was found, completing the proofs.)

But with contradictions inherent in the mathematical system used in a proof, how can one ever really prove anything by contradiction? Imaginary numbers are one example. The square root of $+1$ is a real number because $+1 \times +1 = +1$; however, the square root of -1 is imaginary because -1 times -1 would also equal $+1$, instead of -1. This appears to be a contradiction. Yet it is accepted, and imaginary numbers are used routinely. But how can we justify using them to *prove* a contradiction?

And philosophically speaking, if we can apply inductive logic (and not deductive logic) at any point, who needs a proof in this case? Using inductive logic, F.L.T. is proved after enough examples have been found, and after the last number-crunching effort by computer, few could complain that they haven't.

Deductive logic (as in the Latin *a priori*, "from what comes before") draws specific conclusions from the general case assumed to be true. (For example, the classic syllogism: "All men are mortal. Socrates was a man. Therefore, Socrates was mortal.") Detectives and computers both use deductive logic.

Inductive logic (as in the Latin *a posteriori*, "from what comes after") draws general conclusions from the specific case. (E.g., "The sun has come up every day that has gone before; therefore, it will come up tomorrow.") Scientists routinely use inductive logic. This is the experimental method, and it works very well with, say, biological systems. (That is, we test a new drug thousands of times, and if it's safe and effective, we put it on the market.) But with F.L.T., testing thousands of numbers (millions, actually) *wasn't* good enough to put it on the market.

But the relative merits of deductive versus inductive proofs aside, inductive proofs have relative merits among themselves. To illustrate, let's suppose for a moment that Fermat's last theorem has just been tested with modern hardware for the first time. At your left, there's a powerful computer and a printout of the results, which show the theorem is true for exponents up to four million. At your right, there's a mathematician holding a two-hundred-page document, which concludes that the theorem is true for all numbers. Which do you believe provides the greater certainty?

In short, does anyone believe Fermat's last theorem is true any more now than they did before June of 1993?

[A Possible Fatal Flaw]

A possible fatal flaw in Wiles's proof is whether the same basic arguments could be constructed to hold true for *all* exponents, instead of just the exponents equal to or greater than 3. If it could, the same proof would "prove" the Pythagorean theorem ($x^2 + y^2 = z^2$) to be false.

[A Matter of Credit]

It's also interesting to note the chronology of developments related to the news story and who gets credit for what. The links that make up the chain of proof were forged between

1954 and 1993. But the *chronological* order of the forging of those links is not the same as the *logical* order. Here is the logical order:

[1] Taniyama (in 1954) makes a conjecture.
[2] Wiles (in 1993) offers a proof of a special case of the Taniyama conjecture.
[3] Frey (in 1983) postulates a connection between the Taniyama conjecture and Fermat's last theorem.
[4] Ribet (in 1987) proves the Frey connection.
       Q.E.D.—F.L.T.

This suggests that if the links had followed the logical order instead, it would be Ribet who would now be receiving the credit for proving F. L. T. If Wiles had been faster, he would have forged his link before Ribet's work; if Ribet had been slower, he would have forged his link after Wiles's work. But because these events came out of order chronologically, Wiles received the credit—because he presented his proof last. That is, if the logical sequence of events had come in chronological order, Ribet's work would have been the completing link in the chain, and it would have been *his* name that flashed around the world. So it seems reasonable to say that the two men should now be sharing the credit.

Of course, Wiles did say that he was inspired by Ribet's work, which implies that he might never have worked on his link otherwise, but this inspiration may add as much to the case for Ribet as it subtracts.

[*Why This Proof Wasn't Fermat's Proof*]

But speaking of Fermat again, no one could deny that this modern mathematical proof isn't what he had in mind (if, indeed, he had a proof in mind at all), and few would deny that this proof would have been totally unacceptable to him. (Neither should we make the mistake of crediting Fermat with managing to formulate—and not writing down—a proof that would have compared to Wiles's ingenious, imaginative effort. My own guess is that Fermat thought he had a proof, but later discovered he was mistaken. He was indeed mistaken about another "theorem," which will be detailed in forthcoming pages.)

Wiles' proof depends on many developments in the field of mathematics that were unknown at the time of Fermat and his contemporaries, including many that are recent and poorly understood. Among the great many are Galois theory, modular forms, deformation theory, P-adic numbers, and theories of L-functions.

[*To All the Unknown Genius in the World*]

So we know the proof isn't Fermat's, but that's of less interest than the splendid, undeniable fact that Fermat's last theorem and other problems like it have proved (!) that mathematical thinking has always been fascinating to the world, and the fascination continues unabated to the present day. And many other problems like it remain unproved. (This will continue to be the case throughout time. Conjectures are made. Proofs will follow. It's in the nature of things.) In fact, Fermat himself had one more "theorem" (!), which will be detailed in a very few pages.

Regardless of any individual's particular mathematical goal, if it were my decision to make, I would encourage each and every attempt, both professional *and* amateur. Should people be discouraged from playing chess unless they're members of the United States Chess Federation? Should people be discouraged from exercising unless they're performing in the Olympics? Should we look down upon every old man, every little girl, and every mathematician who runs as hard as he or she can in the New York Marathon?

Mental exercise keeps us fit, and hard thinking does anyone

and everyone a world of good, regardless of whether they win any particular prize. The people who finished dead last in the most recent New York Marathon were just as tired as the front-runners (*more* tired, actually), and they felt wonderful about it. Would we rather that they sit on the sofa and watch the winners cross the finish line on television, instead?

So why do people like Philip Davis and Reuben Hersh get so many letters from oddballs? (Underwood Dudley, who has written two books on crank proofs, calls such oddballs "Fermatists.") In their book *The Mathematical Experience*, they add, "Very often, the correspondent not only 'succeeds' in solving one of the great mathematical unsolvables, but has also found a way to construct an antigravity shield, to interpret the mysteries of the Great Pyramid and of Stonehenge, and is well on his way to producing the Philosophers' Stone. This is no exaggeration." It's probably because the oddballs are usually the only ones who become convinced that they've found the answer and decide to write!

Discovering the next ten-year-old Wiles (who was surely only an "amateur" enthusiast for some years to follow) or the next Ramanujan (see below) isn't the point. The point is that it is honorable (at the very least) to spend the better part of a rainy weekend, whether one or more, bending one's mind to a highly intellectual task. There's a very wide range of people between mathematicians and madmen, and they all need mental exercise.

Srinivasa Ramanujan was surely one of the most intriguing figures in the history of mathematics, unknown to the general public, but recognized by mathematicians as a genius without peer. Unknown, poor, and living in a small town in his native India, he managed to obtain a copy of George Shoobridge Carr's *Synopsis of Elementary Results in Pure and Applied Mathematics* and became utterly absorbed with the subject, even to the extent of developing his own theorems. Later, he managed to secure a scholarship to the University of Madras but lost it because he couldn't concentrate on his other subjects. Jobless and penniless, he continued his work nonetheless, and in 1913, he wrote a letter to G. H. Hardy, one of the leading English mathematicians of the time. Hardy recognized the signs of genius, and came to his aid, even going so far as to bring Romanujan to England and tutor him personally. As Ramanujan had been self-educated, there were great gaps in his knowledge, but his mastery of the subjects he knew was stunning. Eventually, he published papers in English and European mathematics journals, was elected to the Royal Society of London, and contributed pioneering discoveries to number theory.

Hermann Günther Grassman was another unusual individual. An ordained minister in Poland, he turned to mathematics and published a book called *Die Lineale Ausdehnungslehre* in 1844. Although the modern mathematical community regards the book as a work of genius, mathematicians of Grassman's day rejected it outright not only because it was unclear (Grassman invented a symbolism that was, unhappily for everyone, all his own), but also because it was frankly weird and almost mystical.

In their book *The Mathematical Experience*, Philip Davis and Reuben Hersh remind us of the Polish mathematician Jozef Maria Wronski, "whose personality and work combined elements from pretentious naiveté to genius near madness" and who printed his own "key to the universe" on all his works, "placed in a cartouche, sanctified by the zodiac, and guarded by a sphinx." They continue on the subject:

> There is work, then, which is wrong, is acknowledged to be wrong, and which, at some later date, may be set to rights. There is work which is dismissed without examination. There is work which is so obscure that it is difficult to interpret and is perforce ignored. Some of it may emerge later. There is work which may be of great importance . . .

which is heterodox, and as a result, is ignored or boy-
cotted. There is also work, perhaps the bulk of the mathematical output, which is admittedly correct, but which in
the long run is ignored for lack of interest or because the
main streams of mathematics did not choose to pass that
way. In the final analysis, there can be no formalization of
what is right and how we know it is right, what is accepted,
and what the mechanism for acceptance is. As Hermann
Weyl has written, "Mathematizing may well be a creative
activity of man . . . whose historical decisions defy complete objective rationalization."

But if you know you're not a Ramanujan or a Grassman or a
Wronski, you should be aware of the following. In *Wheels, Life
and Other Mathematical Amusements*, Martin Gardner notes:

> The mathematics departments of many large universities
> return all proofs of Fermat's last theorem with a form letter
> stating that the paper will be evaluated only after an ad-
> vance payment of a specified fee. Edmund Landau, a Ger-
> man mathematician, used a form letter that read: "Dear
> Sir/ Madam: Your proof of Fermat's last theorem has been
> received. The first mistake is on page ____ , line _____."
> Landau would then assign the filling in of the blanks to a
> graduate student.

[The Search for Other Proofs]

Before we get back to Fermat, you might want to consider
performing the following mathematical feats:

[1] Wiles proved F.L.T. with concepts from hyperbolic (non-Euclidean) geometry in the way that Bolyai squared the circle in
hyperbolic geometry, a construction that was proven impossible in abstract algebra. Use Euclidean geometry or abstract
algebra to prove that F.L.T. is impossible to prove.

[2] Better yet, use Euclidean geometry to prove that F.L.T. is true, which will be more of an achievement than a non-Euclidean proof. Or use Euclidean geometry to prove that F.L.T. is true with a direct proof. (This is guaranteed to thrill every formalist on the planet and can hardly be surpassed.)

[3] And for the most ambitious, use hyperbolic geometry to prove that F.L.T. is impossible to prove. A contradiction of this magnitude (and fame) could cause the entire field to collapse. (But don't expect any of the experts in hyperbolic geometry to want to verify your proof.)

[4] To demolish Einstein's theories of relativity (in elliptic geometry), you could go back and prove the parallel postulate, bringing down not only Einstein, but all of the non-Euclidean geometries, as well. (Or you could go after Einstein selectively by the route of proving a contradiction in elliptic geometry.)

[*Fermat's Other Unproved Theorem*]

But if you don't want to spend the rest of your life trying to give all modern mathematicians a nervous breakdown, you could narrow your enthusiasm to Fermat's *other* unproved "theorem." Fermat, you see, made mistakes just like the rest of us. One of his unproved conjectures turned out *false* a century later. (Given everything you've read in this book, it's a little unsettling to think about, isn't it?)

Back in 1640, three years *after* what we now call Fermat's "last" theorem was written in a marginal note, he wrote to several of his friends and stated that all numbers of the construction $2^{2^n} + 1$ are prime. (Prime numbers of this construction are now known as "Fermat primes".) However, Euler showed a century afterward that the number $2^{2^5} + 1$ has a factor of 641. (That is, it can be divided evenly by 641.) This means it is not a prime number, so Fermat was wrong. It is unknown if there are any Fermat primes above Euler's non-prime number of 4,294,967,297. (2^{2^5} + 1 = 2^{32} + 1 = 4,294,967,296 + 1 = 4,294,967,297. Because 4,294,967,297 can be divided evenly by 641, it is not a prime number, the way Fermat mistakenly conjectured.)

Clearly, an inductive proof would have been a failure here, at least in the days before computers. And even with computers on the scene, who's to say when enough searching is enough? An exception turning up in the neighborhood of 4,294,967,297 must surely have been a surprise to Euler. Considering how many arithmetic tools have been created expressly for the purpose of proving F.L.T.—and not disproving it—we'd better hope that Fermat's last theorem is indeed true to begin with and that there isn't an exception sitting just beyond the current horizon the way there was before.

The question now is, do there exist infinitely many Fermat primes of the construction 2^{2^n} + 1? This way, only *you* are likely to have the nervous breakdown (regardless of whether you succeed).

In the book *An Introduction to Mathematics*, the famed Alfred North Whitehead wrote, "The study of mathematics is apt to commence in disappointment . . . We are told that by its aid the stars are weighed, and the billions of molecules in a drop of water are counted. Yet, like the ghost of Hamlet's father, this great science eludes the efforts of our mental weapons to grasp it."

If Wiles's proof holds up in "math court," his place in the history of mathematics is assured. But that's because he solved a famous problem. That shouldn't make him outshine the many, many more mathematicians out there working quietly away, day after day, honorably and productively, at far less famous tasks than his. They, along with many others, are responsible for much of the quality of life we Americans are now able to take for granted, from videotape recorders to computers to communications satellites.

And if Andrew Wiles really did have that keen sense of humor "postulated" in the first part of this book, you know what he'd do, don't you? In order to bedevil future generations of mathematicians, he'd leave a little note somewhere in the margin of his copy of Albert Einstein's last paper entitled "Relativistic Theory of the Non-Symmetric Field" to wit:

> I now know that this is impossible, and I've found a remarkable proof, but the margin is too small to contain it.

A Poem by Cody Pfanstiehl

X to the nth plus Y to the nth
Equals Z to the nth. This is true,
But it's only when
The number for n
Is a number that's not more than 2.

Yes, the square of the X and the square of the Y
Does equal the square of the Z,
If you limit the exponents
On the left side
To superscripts smaller than 3.

Put your X in one basket but please do not ask it
To raise itself more than a deuce.
And you would be wise
To limit your Ys
To a number not greater than tuece.

A 3 raised to 9 is just dandy and fine
Plus a 4 to 16 is real cool.
But larger than that
All the sums are too fat
As surely they taught you in school.

A 3 times a 3 plus a 4 times a 4
Add up to a 5 times a 5,
But is there a proof
That is not just a spoof
That exponents larger survive?

Who can jostle and jigger the numbers with vigor
To prove Pierre Fermat's old theorem?
Through ages the sages
Wrote thousands of pages
But nobody came even near 'im

Until Andy Wiles who presented, with smiles,
A 200 page affirmation.
He accepted the praise.
Now he hopes for a raise
Which is better than mere adulation.

His colleagues all cheered (save a few who were weird)
As they set out to check out his math.
Now they'll ask him, these rubes,
To divide up those cubes
And we'll see who will have the last laugh.

Appendix

[*Karl Rubin's Highlights of Andrew Wiles's Proof*]

Hello netters, local number theorist [no crank . . . he's the most recent recipient of the AMS Cole Prize in Number theory] Prof. Karl Rubin was present at the Wiles lectures in Cambridge. He has posted the following outline of the proof to our math newsgroup:

From K. C. Rubin@newton.cam.ac.uk Thu Jun 24 14:50:52 EDT 1993
Article: 535 of math.announce
Path: math.ohio-state.edu!gateway
From: K.C.Rubin@newton.cam.ac.uk
Newsgroups: math.announce
Subject: sketch of Fermat
Date: 24 June 1993 09:19:10 -0400

Organization: The Ohio State University, Department
  of Mathematics
Lines: 103
Sender: daemon@math.ohio-state.edu
Message-ID: <m0o8rAP-
  00005sC@newton.newton.cam.ac.uk>
NNTP-Posting-Host: mathserv.mps.ohio-state.edu

Several people have asked for more details about
Andrew's proof. Here is a lengthy sketch. Enjoy.

Karl

Theorem. *If E is a semistable elliptic curve defined over* $\mathbf{Q}$*, then
E is modular.*

It has been known for some time, by work of Frey and Ribet, that Fermat follows from this. If $u^q + v^q + w^q = 0$, then
Frey had the idea of looking at the (semistable) elliptic curve
$y^2 = x(x - u^q)(x + v^q)$. If this elliptic curve comes from a modular form, then the work of Ribet on Serre's conjecture shows
that there would have to exist a modular form of weight 2 on
$\Gamma_0(2)$. But there are no such forms.

To prove the Theorem, start with an elliptic curve E, a
prime p and let

$$\rho_p \colon \mathrm{Gal}\,(\bar{\mathbf{Q}}/\mathbf{Q}) \to \mathrm{GL}_2\,(\mathbf{Z}/\mathbf{pZ})$$

be the representation giving the action of Galois on the p-
torsion $E[p]$. We wish to show that a *certain* lift of this representation to $\mathrm{GL}_2(\mathbf{Z}_p)$ (namely, the p-adic representation on
the Tate module $T_p(E)$) is attached to a modular form. We will
do this by using Mazur's theory of deformations, to show that
every lifting which 'looks modular' in a certain precise sense is
attached to a modular form.

Fix certain 'lifting data', such as the allowed ramification,
specified local behavior at p, etc. for the lift. This defines a

lifting problem, and Mazur proves that there is a universal lift, i.e. a local ring R and a representation into $GL_2(R)$ such that every lift of the appropriate type factors through this one.

Now suppose that ρ_p is modular, i.e. there is *some* lift of ρ_p which is attached to a modular form. Then there is also a Hecke ring T, which is the maximal quotient of R with the property that all *modular* lifts factor through T. It is a conjecture of Mazur that $R = T$, and it would follow from this that *every* lift of ρ_p which "looks modular" (in particular the one we are interested in) is attached to a modular form.

Thus we need to know 2 things:

(a) ρ_p is modular
(b) $R = T$.

It was proved by Tunnell that ρ_3 is modular for every elliptic curve. This is because $PGL_2(\mathbf{Z/3Z}) = S_4$. So (a) will be satisfied if we take $p = 3$. This is crucial.

Wiles uses (a) to prove (b) under some restrictions on ρ_p. Using (a) and some commutative algebra (using the fact that T is Gorenstein, "basically due to Mazur") Wiles reduces the statement $T = R$ to checking an inequality between the sizes of 2 groups. One of these is related to the Selmer group of the symmetric square of the given modular lifting of ρ_p, and the other is related (by work of Hida) to an L-value. The required inequality, which everyone presumes is an instance of the Block-Kato conjecture, is what Wiles needs to verify.

He does this using a Kolyvagin-type Euler system argument. This is the most technically difficult part of the proof, and is responsible for most of the length of the manuscript. He uses modular units to construct what he calls a 'geometric Euler system' of cohomology classes. The inspiration for his construction comes from work of Flach, who came up with what is essentially the 'bottom level' of this Euler system. But Wiles needed to go much farther than Flach did. In the end, *under certain hypotheses* on ρ_p he gets a workable Euler system and proves the desired inequality. Among other things, it is necessary that ρ_p is irreducible.

Suppose now that E is semistable.

Case 1. ρ_3 is irreducible.

Take $p=3$. By Tunnell's theorem (a) above is true. Under these hypotheses the argument above works for ρ_3, so we conclude that E is modular.

Case 2. ρ_3 is reducible.

Take $p=5$. In this case ρ_5 must be irreducible, or else E would correspond to a rational point on $X_0(15)$. But $X_0(15)$ has only 4 noncuspidal rational points, and these correspond to non-semistable curves. *If* we knew that ρ_5 were modular, then the computation above would apply and E would be modular.

We will find a new semistable elliptic curve E' such that $\rho_{E,5}$ = $\rho_{E',5}$ and $\rho_{E',3}$ is irreducible. Then by Case I, E' is modular. Therefore $\rho_{E,5} = \rho_{E',5}$ does have a modular lifting and we will be done.

We need to construct such an E'. Let X denote the modular curve whose points correspond to pairs (A, C) where A *is* an elliptic curve and C is a subgroup of A isomorphic to the group scheme $E[5]$. (All such curves will have mod-5 representation equal to ρ_E.) This X is genus 0, and has one rational point corresponding to E, so it has infinitely many. Now Wiles uses a Hilbert Irreducibility argument to show that not all rational points can be images of rational points on modular curves covering X, corresponding to degenerate level 3 structure (i.e. $\mathrm{im}(\rho_3) \neq GL_2(\mathbf{Z}/3)$). In other words, an E' of the type we need exists. (To make sure E' is semistable, choose it 5-adically close to E. Then it is semistable at 5, and at other primes because $\rho_{E',5} = \rho_{E,5}$.)

For Further Reading

Bell, Eric Temple, *The Last Problem*, New York: Simon & Schuster, 1961.

Davis, Philip and William G. Chinn, *3.1416 and All That*, Boston: Birkhäuser Boston, 1985.

Davis, Philip J., and Reuben Hersh, *The Mathematical Experience*, Boston: Birkhäuser Boston, 1981.

Eddington, Sir Arthur, *The Expanding Universe*, Cambridge: Cambridge University Press, 1933.

Edwards, Harold, *Fermat's Last Theorem*, New York: Springer-Verlag, 1977.

Gardner, Martin, *Annotated Alice: Alice's Adventures in Wonderland & Through the Looking Glass*, New York: NAL-Dutton, 1974.

Gardner, Martin, *Wheels, Life and Other Mathematical Amusements*, New York: W. H. Freeman & Company, 1983.

Gray, Jeremy, *Ideas of Space*, Oxford: Oxford University Press, 1989.

Holten, G., *Thematic Origins of Scientific Thought, Kepler to Einstein*, Cambridge, MA: Harvard University Press, 1973.

Kanigel, Robert, *A Man Who Knew Infinity* (biography of Srinivasa Ramanujan), New York: Scribners, 1991.

Jones, Arthur, Sidney A. Morris, Kenneth R. Pearson, *Abstract Algebra and Famous Impossibilities*, New York: Springer-Verlag, 1991.

Kline, Morris, *Mathematics for the Non-Mathematician*, New York: Dover Publications, 1967.

Martin, George E., *The Foundations of Geometry and the Non-Euclidean Plane*, New York: Springer-Verlag, 1982.

Meschkowski, H., *The Evolution of Mathematical Thought*, New York: Holden-Day, 1965.

Ribenboim, Paulo, *Thirteen Lectures on Fermat's Last Theorem*, New York: Springer-Verlag, 1979.

Salmon, Wesley, editor, *Zeno's Paradoxes*, New York: Bobbs-Merrill Educational Publishing, 1970.

Ulam, Stanislaw, *Adventures of a Mathematician*, New York: Scribners, 1976.

Weil, André, *Number Theory*, Boston: Birkhäuser Boston, 1984.

Wilder, Raymond L., *The Evolution of Mathematical Concepts*, New York: John Wiley & Sons, 1968.

About the Author

MARILYN VOS SAVANT was born in St. Louis, Missouri, the daughter of Mary vos Savant and Joseph Mach. She is married to Robert Jarvik, M.D., inventor of the Jarvik 7 artificial heart. They live in New York City.

She was listed in the *Guinness Book of World Records* for five years under "Highest I.Q." for both childhood and adult scores, and has now been inducted into the Guinness Hall of Fame. She is a writer, lecturer, and spends additional time assisting her husband in the artificial heart program. Her special interests and concerns are quality education and thinking in America, and humanitarian medicine and research. She describes herself as an "independent" with regard to politics and religion, and only an "armchair" feminist.

Marilyn vos Savant writes the "Ask Marilyn" question-and-answer problem-analysis column for *Parade*, the Sunday magazine for 352 newspapers, with a circulation of 36 million and a readership of 70 million, the largest in the world. Her *Ask Marilyn* trade paperback will be published in September, 1993, her *"I've Forgotten Everything I Learned in School!"* hardcover will be published in January, 1994, and her *Number Blindness* hardcover will be published in September, 1994, all by St. Martin's Press. Her *Ask Marilyn* hardcover was published by St. Martin's Press in 1992.

Her stage play called *It Was Poppa's Will* was produced in staged reading, and she has also written a fantasy/satire (novel) of a dozen classical civilizations in history called *The Re-Creation* and a futuristic political fantasy/satire (novel), as yet untitled.

But far from the stereotype of the intellectual, vos Savant says she believes that "an ounce of sequins is worth a pound of home cooking," and doesn't engage in the latter "for humanitarian reasons." And what does she do for fun? Read a book? Retreat to the wilderness? She looks surprised. "Not at all. My idea of fun is going out with people," she says. "A park is a nice place to visit, but I wouldn't want to live there. I'd rather be surrounded by a thousand people than a thousand trees." Her hobby is writing letters to friends around the world.